Student's Solutions Manual

for

Modern Physics
Fifth Edition

Paul A. Tipler
Ralph A. Llewellyn

Prepared by
Mark J. Llewellyn
School of Electrical Engineering and Computer Science
University of Central Florida

W. H. Freeman and Company
New York

Printed in the United States of America

ISBN-13: 978-0-7167-8575-3
ISBN-10: 0-7167-8475-0

First printing

W. H. Freeman and Company
41 Madison Avenue
New York, NY 10010
Houndmills, Basingstoke
RG21 6XS, England
www.whfreeman.com

Table of Contents

Preface

This book is the Student's Solutions Manual for the end-of-chapter problems that appear in *Modern Physics, Fifth Edition,* by Paul A. Tipler and Ralph A. Llewellyn. This book contains solutions to every fourth end-of-chapter problem in the text.

Figure numbers, equations, and table numbers refer to those in the text. Figures in this solutions manual are not numbered and correspond only to the problem in which they appear. Notation and units parallel those in the text.

Please visit the W. H. Freeman Physics Web site at www.whfreeman.com/tiplermodernphysics5e where you will find 30 MORE sections that expand on high-interest topics covered in the textbook; the Classical Concept Review, which provides refreshers for many classical physics topics that are background for modern physics topics in the text; and an image gallery for Chapter 13. Some problems in the text are drawn from the MORE sections.

Every effort has been made to ensure that the solutions in this manual are accurate and free from errors. If you find an error or a better solution to any of these problems, please feel free to contact me at the address below with a specific citation. I appreciate any correspondence from users of this manual who have ideas and suggestions for improving it.

Mark J. Llewellyn
School of Electrical Engineering and Computer Science
University of Central Florida
Orlando, Florida 32816-2362
Email: markl@cs.ucf.edu

Chapter 1 – Relativity I

1-1.   (a) Speed of the droid relative to Hoth, according to Galilean relativity, u_{Hoth}, is

$$u_{Hoth} = u_{spaceship} + u_{droid}$$
$$= 2.3 \times 10^8 \, m/s + 2.1 \times 10^8 \, m/s$$
$$= 4.4 \times 10^8 \, m/s$$

(b) No, since the droid is moving faster than light speed relative to Hoth.

1-5.   (a) In this case, the situation is analogous to Example 1-1 with $L = 3 \times 10^8 \, m$,

$v = 3 \times 10^4 \, m/s$, and $c = 3 \times 10^8 \, m/s$ If the flash occurs at $t = 0$, the interior is dark

until $t = 2s$ at which time a bright circle of light reflected from the circumference of

the great circle plane perpendicular to the direction of motion reaches the center, the

circle splits in two, one circle moving toward the front and the other moving toward

the rear, their radii decreasing to just a point when they reach the axis $10^{-8} \, s$ after

arrival of the first reflected light ring. Then the interior is dark again.

(b) In the frame of the seated observer, the spherical wave expands outward at c in all

directions. The interior is dark until $t = 2s$ at which time the spherical wave (that

reflected from the inner surface at $t = 1s$) returns to the center showing the entire inner

surface of the sphere in reflected light, following which the interior is dark again.

1-9.   The wave from the front travels 500 m at speed $c + (150/3.6)$ m/s and the wave from the

rear travels at $c - (150/3.6)$ m/s. As seen in Figure 1-14, the travel time is longer for the

wave from the rear.

$$\Delta t = t_r - t_f = \frac{500m}{3.00 \times 10^8 \, m/s - (150/3.6) m/s} - \frac{500m}{3.00 \times 10^8 \, m/s + (150/3.6) m/s}$$

$$= 500 \left[\frac{3 \times 10^8 + (150/3.6) - 3 \times 10^8 + (150/3.6)}{(3 \times 10^8)^2 - 2(150/3.6)(3 \times 10^8) - (150/3.6)^2} \right]$$

(Problem 1-9 continued)

$$= 500\frac{2(150/3.6)}{(3\times10^8)^2} \approx 4.63\times10^{-13}s$$

1-13. (a) $\gamma = 1/(1-v^2/c^2)^{1/2} = 1/\left[1-(0.85c)^2/c^2\right]^{1/2} = 1.898$

$x' = \gamma(x-vt) = 1.898\left[75m-(0.85c)(2.0\times10^{-5}s)\right] = -9.537\times10^3 m$

$y' = y = 18m$

$z' = z = 4.0m$

$t' = \gamma(t-vx/c^2) = 1.898\left[2.0\times10^{-5}s-(0.85c)(75m)/c^2\right] = 3.756\times10^{-5}s$

(b) $x = \gamma(x'+vt') = 1.898\left[-9.537\times10^3 m+(0.85c)(3.756\times10^{-5}s)\right] = 75.8m$

difference is due to rounding of γ, x', and t'.

$y = y' = 18m$

$z = z' = 4.0m$

$t = \gamma(t'+vx'/c^2) = 1.898\left[3.756\times10^{-5}s+(0.85c)(-9.537\times10^3 m)/c^2\right] = 2.0\times10^{-5}s$

1-17. (a) As seen from the diagram, when the observer in the rocket (S') system sees 1 $c\cdot s$ tick by on the rocket's clock, only 0.6 $c\cdot s$ have ticked by on the laboratory clock.

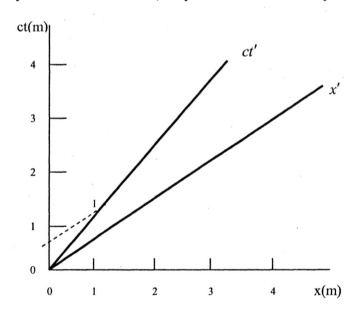

(b) When 10 seconds have passed on the rocket's clock, only 6 seconds have passed on the laboratory clock.

2

1-21.

$$\Delta t = \gamma \Delta t' \qquad \text{(Equation 1-26)}$$

$$\frac{\Delta t - \Delta t'}{\Delta t'} = \frac{\gamma \Delta t' - \Delta t'}{\Delta t'} = \gamma - 1 \approx \frac{1}{2}\frac{v^2}{c^2} = \gamma - 1 \approx \frac{1}{2}\frac{v^2}{c^2}$$

$$v^2 = 2c^2 \frac{\Delta t - \Delta t'}{\Delta t'} \qquad v = c\left(2 \times \frac{\Delta t - \Delta t'}{\Delta t'}\right)^{1/2} = c\left(2 \times 0.01\right)^{1/2} = 0.14c$$

1-25. From Equation 1-28, $L = L_p / \gamma = L_p \sqrt{1 - v^2/c^2}$ where $L = 85m$ and $L_p = 100m$

$$\sqrt{1 - v^2/c^2} = L/L_p = 85/100$$

Squaring $\quad 1 - v^2/c^2 = \left(85/100\right)^2$

$$\therefore \quad v^2 = \left[1 - \left(85/100\right)^2\right]c^2 = 0.2775c^2 \text{ and } v = 0.527c = 1.58 \times 10^8 \, m/s$$

1-29. (a) In $S': V' = a' \times b' \times c' = (2m)(2m)(4m) = 16m^3$

In S: Both a' and c' have components in the x' direction.

$a'_x = a' \sin 25° = (2m)\sin 25° = 0.84m$ and $c'_x = c' \cos 25° = (4m)\cos 25° = 3.63m$

$a_x = a'_x \sqrt{1 - \beta^2} = 0.84\sqrt{1 - (0.65)^2} = 0.64m$

$c_x = c'_x \sqrt{1 - \beta^2} = 3.634\sqrt{1 - (0.65)^2} = 2.76m$

$a_y = a'_y = a' \cos 25° = 2\cos 25° = 1.81m$ and $c_y = c'_y = c' \sin 25° = 4\sin 25° = 1.69m$

$a = \sqrt{a_x^2 + a_y^2} = \sqrt{(0.64)^2 + (1.81)^2} = 1.92m$

$c = \sqrt{c_x^2 + c_y^2} = \sqrt{(2.76)^2 + (1.69)^2} = 3.24m$

b' (in z direction) is unchanged, so $b = b' = 2m$

θ (between c and xy-plane) $= \tan^{-1}(1.69/2.76) = 31.5°$

ϕ (between a and yz-plane) $= \tan^{-1}(0.64/1.81) = 19.5°$

$V = $ (area of ay face) $\bullet\ b$ (see part[b])

$\qquad V = (c \times a \sin 78°) \times b = (3.24m)(1.92m\ \sin 78°)(2m) = 12.2m^3$

(Problem 1-29 continued)

(b)

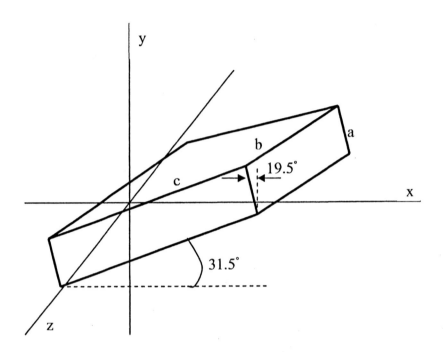

$$\lambda' = \frac{c}{f'} = \frac{c}{f_o\sqrt{\dfrac{1+\beta}{1-\beta}}} = \sqrt{\frac{1-v/c}{1+v/c}}\,\lambda_o \quad \text{(Equation 1-36)}$$

$$\left(\frac{\lambda'}{\lambda_o}\right)^2 = \frac{1-v/c}{1+v/c} \quad \rightarrow \quad \left(\frac{\lambda'}{\lambda_o}\right)^2 (1+v/c) = 1-v/c$$

1-33.

$$f = \sqrt{\frac{1-\beta}{1+\beta}}\,f_o \quad \rightarrow \quad \lambda = \sqrt{\frac{1-\beta}{1+\beta}}\,\lambda_o = \sqrt{\frac{1-\beta}{1+\beta}}\,(656.3nm)$$

For $\beta = 10^{-3}$: $\quad \lambda = (656.3nm)\sqrt{\dfrac{1+10^{-3}}{1-10^{-3}}} = 657.0nm$

For $\beta = 10^{-2}$: $\quad \lambda = (656.3nm)\sqrt{\dfrac{1+10^{-2}}{1-10^{-2}}} = 662.9nm$

$\beta = 10^{-1}$: $\quad \lambda = (656.3nm)\sqrt{\dfrac{1+10^{-1}}{1-10^{-1}}} = 725.6nm$

1-37.
$$\cos\theta = \frac{\cos\theta' + \beta}{1 + \beta\cos\theta'} \quad \text{(Equation 1-41)}$$

where θ' = half-angle of the beam in $S' = 30°$

For $\beta = 0.65$, $\cos\theta = \dfrac{\cos 30° + 0.65}{1 + (0.65)\cos 30°} = 0.97$ or $\theta = 14.1°$

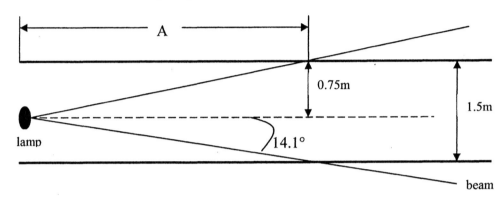

The train is A from you when the headlight disappears, where $A = \dfrac{0.75m}{\tan 14.1°} = 3.0m$

1-41. Orbit circumference $= 4.0 \times 10^7 m$.

Satellite speed $v = 4.0 \times 10^7 m / (90\min \times 60s/\min) = 7.41 \times 10^3 m/s$

$$\Delta t - \Delta t_0 = t_{diff}$$

$$\Delta t - \Delta t/\gamma = t_{diff} = \Delta t(1 - 1/\gamma) = \Delta t\left(\frac{1}{2}\beta^2\right) \quad \text{(Problem 1-20)}$$

$$t_{diff} = \left(3.16 \times 10^7 m/s\right)(1/2)\left(7.41 \times 10^3 / 3.0 \times 10^8\right)^2$$
$$= 0.0096s = 9.6ms$$

1-45. (a)

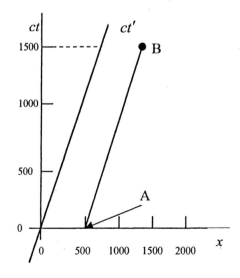

(Problem 1-45 continued)

(b) Slope of ct' axis $= 2.08 = 1/\beta$, so $\beta = 0.48$ and $v = 1.44 \times 10^8 \, m/s$

(c) $ct' = \gamma ct$ and $\gamma = 1/\sqrt{1-\beta^2}$ so $ct'\sqrt{1-\beta^2} = ct$

For $ct' = 1000m$ and $\beta = 0.48$ $\qquad ct = 877m \quad t' = 1.5(877)/c = 4.39 \mu s$

(d) $\Delta t = \gamma \Delta t' = 1.14\Delta t' \quad \rightarrow \quad \Delta t' = 5\mu s/1.14 = 4.39 \mu s$

1-49. $t_2' - t_1' = \gamma(t_2 - t_1) - \dfrac{\gamma v}{c^2}(x_b - x_a)$ (Equation 1-20)

(a) $t_2' - t_1' = 0 \quad \rightarrow \quad (t_2 - t_1) = (v/c^2)(x_b - x_a) \quad \rightarrow \quad (0.5-1.0)y = (v/c^2)(2.0-1.0)c \cdot y$

Thus, $-0.5 = (v/c) \quad \rightarrow \quad v = 0.5c$ in the $-x$ direction.

(b) $t' = \gamma(t - vx/c^2)$

Using the first event to calculate t' (because t' is the same for both events),

$$t = \left(1/\sqrt{1-(0.5)^2}\right)\left[1y - (0.5c)(1c \cdot y)/c^2\right] = 1.155(0.5) = 0.58y$$

(c) $(\Delta s)^2 = (\Delta x)^2 - (c\Delta t)^2 = (1c \cdot y)^2 - (0.5c \cdot y)^2 = 0.75(c \cdot y)^2 \quad \rightarrow \quad \Delta s = 0.866c \cdot y$

(d) The interval is spacelike.

(e) $L = \Delta s = 0.866c \cdot y$

1-53. This is easier to do in the xy and $x'y'$ planes. Let the center of the meterstick, which is parallel to the x-axis and moves upward with speed v_y in S, at $x = y = x' = y' = 0$ at $t = t' = 0$. The right hand end of the stick, e.g., will not be at $t' = 0$ in S' because the clocks in S' are not synchronized with those in S. In S' the components of the sticks velocity are:

$$u_y' = \frac{u_y}{\gamma\left(1 - vu_x/c^2\right)} = \frac{v_y}{\gamma} \text{ because } u_y = v_y \text{ and } u_x = 0$$

$$u_x' = \frac{u_x - v}{1 - vu_x/c^2} = -v \text{ because } u_x = 0$$

6

(Problem 1-53 continued)

When the center of the stick is located as noted above, the right end in S' will be at:

$x' = \gamma(x - vt) = 0.5\gamma$ because $t = 0$. The S' clock there will read: $t' = \gamma(t - vx/c^2) = -0.5\gamma v/c^2$

because $t = 0$. Therefore, when $t' = 0$ at the center, the right end is at $x'y'$ given by:

$$x' = 0.5\gamma \qquad y' = u_y' t' = \frac{v_y}{v}\left(\frac{0.5\gamma v}{c^2}\right)$$

and $\theta' = \tan^{-1}\dfrac{y'}{x'} = \dfrac{v_y}{\gamma}\left(\dfrac{0.5\gamma v}{c^2}\right)/0.5\gamma = \left(v_y v/c^2\right)\sqrt{1-\beta^2}$

For $\beta = 0.65 \qquad \theta' = \left(0.494 v_y/c\right)$

1-57. $v = 0.6c \qquad \gamma = \dfrac{1}{\sqrt{1-(0.6)^2}} = 1.25$

(a) The clock in S reads $\gamma \times 60\,\text{min} = 75\,\text{min}$ when the S' clock reads 60 min and the first signal from S' is sent. At that time, the S' observer is at $v \times 75\,\text{min} = 0.6c \times 75\,\text{min} = 45c\cdot\text{min}$. The signal travels for 45 min to reach the S observer and arrives at 75 min + 45 min = 120 min on the S clock.

(b) The observer in S sends his first signal at 60 min and its subsequent wavefront is found at $x = c(t - 60\,\text{min})$. The S' observer is at $x = vt = 0.6ct$ and receives the wavefront when these x positions coincide, i.e., when

$c(t - 60\,\text{min}) = 0.6ct$

$0.4ct = 60c\cdot\text{min}$

$t = (60c\cdot\text{min})/0.4c = 150\,\text{min}$

$x = 0.6c(0\,\text{min}) = 90c\cdot\text{min}$

The confirmation signal sent by the S' observer is sent at that time and place, taking 90 min to reach the observer in S. It arrives at 150 min + 90 min = 240 min.

(c) Observer in S:

Sends first signal	60 min
Receives first signal	120 min
Receives confirmation	240 min

The S' observer makes identical observations.

1-61.   (a) Apparent time $A \rightarrow B = T/2 - t_A + t_B$ and apparent time $B \rightarrow A = T/2 + t_A - t_B$ where

t_A = light travel time from point A to Earth and t_B = light travel time from point B to Earth.

$$A \rightarrow B = \frac{T}{2} - \frac{L}{c+v} + \frac{L}{c-v} = \frac{T}{2} + \frac{2vL}{c^2 - v^2}$$

$$B \rightarrow A = \frac{T}{2} - \frac{L}{c+v} - \frac{L}{c-v} = \frac{T}{2} - \frac{2vL}{c^2 - v^2}$$

(b) Star will appear at A and B simultaneously when $t_B = T/2 + t_A$ or when the period is:

$$T = 2[t_B - t_A] = 2\left[\frac{L}{c-v} - \frac{L}{c+v}\right] = \frac{4vL}{c^2 - v^2}$$

Chapter 2 – Relativity II

2-1. $u_{yB}^2 = u_o^2 \left(1 - v^2/c^2\right)$ $u_{xB}^2 = v^2$

$$\sqrt{1 - \left(u_{xB}^2 + u_B^2\right)/c^2} = \sqrt{1 - v^2/c^2 - \left(u_o^2/c^2\right)\left(1 - v^2/c^2\right)}$$

$$= \sqrt{\left(1 - v^2/c^2\right)\left(1 - u_o^2/c^2\right)}$$

$$= \left(1 - v^2/c^2\right)^{1/2} \left(1 - u_o^2/c^2\right)^{1/2}$$

$$p_{yB} = \frac{mu_{yB}}{\sqrt{1 - \left(u_{xB}^2 + u_{yB}^2\right)/c^2}} = \frac{-mu_o\sqrt{1 - v^2/c^2}}{\sqrt{1 - v^2/c^2}\sqrt{1 - u_o^2/c^2}}$$

$$= -mu_o / \sqrt{1 - u_o^2/c^2} = -p_{yA}$$

2-5. $\Delta E = \Delta mc^2$ $\therefore$ $\Delta m = \Delta E/c^2 = \dfrac{10J}{\left(3.08 \times 10^8 m/s\right)^2} = 1.1 \times 10^{-16} kg$

Because work is done *on* the system, the mass *increases* by this amount.

2-9. $E = \gamma mc^2$ (Equation 2-10)

(a) $200 GeV = \gamma\left(0.938 GeV\right)$ where $mc^2 \left(\text{proton}\right) = 0.938 GeV$

$$\gamma = \frac{1}{\sqrt{1 - v^2/c^2}} = \frac{200 GeV}{0.938 GeV} = 213$$

$$\frac{v}{c} \approx 1 - \frac{1}{2\gamma^2}$$ (Equation 2-40)

$$\frac{v}{c} = 1 - \frac{1}{2\left(213\right)^2} = 1 - 0.00001102 \text{ thus, } v = 0.99998898c$$

(b) $E \approx pc$ for $E \gg mc^2$ where $E = 200 GeV \times 197 = 3.04 \times 10^4 GeV$ (Equation 2-36)

$$p = E/c = 3.94 \times 10^4 GeV/c$$

(Problem 2-9 continued)

(c) Assuming one *Au* nucleus (system S') to be moving in the $+x$ direction of the lab (system *S*), then *u* for the second *Au* nucleus is in the $-x$ direction. The second *Au's* energy measured in the S' system is:

$$E' = \gamma\left(E + vp_x\right) = \left(4.20 \times 10^4\right)\left(3.94 \times 10^4\, GeV + v3.94 \times 10^4\, GeV/c\right)$$

$$= \left(4.20 \times 10^4\right)\left(3.94 \times 10^4\, GeV\right)\left(1 + v/c\right)$$

$$= \left(4.20 \times 10^4\right)\left(3.94 \times 10^4\, GeV\right)\left(2\right)$$

$$= 3.31 \times 10^9\, GeV$$

$$p'_x = \gamma\left(p_x - vE/c^2\right) = \left(4.20 \times 10^4\right)\left(-3.94 \times 10^4\, GeV - v3.94 \times 10^4\, GeV/c^2\right)$$

$$= -\left(4.20 \times 10^4\right)\left(3.94 \times 10^4\, GeV\right)\left(2\right)$$

$$= -3.31 \times 10^9\, GeV/c$$

2-13. (a)

$$p_R = \gamma m_S v \qquad p_N = m_S v$$

$$\gamma = 1/\sqrt{1 - v^2/c^2} = 1/\sqrt{1 - \left(2.5 \times 10^5/3.0 \times 10^8\right)^2}$$

$$\gamma = 1.00000035$$

$$\frac{p_R - p_N}{p_R} = \frac{\gamma m_S\left(2.5 \times 10^3\right) - m_S\left(2.5 \times 10^3\right)}{\gamma m_S\left(2.5 \times 10^3\right)} = \frac{\gamma - 1}{\gamma} = \frac{3.5 \times 10^{-7}}{1.00000035} = 3.5 \times 10^{-7}$$

(b)

$$E_R = \gamma m_S c^2 - m_S c^2 = \left(\gamma - 1\right) m_S c^2$$

$$E_N = \frac{1}{2} m_S v^2$$

$$\frac{E_R - E_N}{E_R} = \frac{\left(\gamma - 1\right)c^2 - 0.5v^2}{\left(\gamma - 1\right)c^2}$$

$$= \frac{3.5 \times 10^{-7} c^2 - 0.5v^2}{3.5 \times 10^{-7} c^2}$$

$$= 1 - \frac{0.5\left(2.5 \times 10^5\right)^2}{3.5 \times 10^{-7} c^2} = 0.0079$$

2-17. $^3H \rightarrow {}^2H + n$

Energy to remove the $n = 22.014102u\left({}^2H\right) + 1.008665u\left(n\right) - 3.016049u\left({}^3H\right)$

$= 0.006718u \times 931.5MeV/u = 6.26MeV$

2-21. Conservation of energy requires that $E_i^2 = E_f^2$, or

$$\left(p_ic\right)^2 + \left(2m_pc^2\right)^2 = \left(p_fc\right)^2 + \left(2m_pc^2 + m_\pi c^2\right)^2 \text{ and conservation of momentum requires that}$$

$p_i = p_f$, so

$$4\left(m_pc^2\right)^2 = 4\left(m_pc^2\right)^2 + 2m_pc^2 \times 2m_\pi c^2 + \left(m_\pi c^2\right)^2$$

$$0 = 2m_pc^2 \times 2m_\pi c^2 + \left(m_\pi c^2\right)^2$$

$$0 = m_\pi c^2\left(2 + \frac{m_\pi c^2}{2m_pc^2}\right) = m_\pi c^2\left(2 + \frac{m_\pi}{2m_p}\right)$$

Thus, $m_\pi c^2\left(2 + m_\pi/2m_p\right)$ is the minimum or threshold energy E_i that a beam proton must have to produce a π^0.

$$E = m_\pi c^2\left(2 + \frac{m_\pi c^2}{2m_pc^2}\right) = 135MeV\left(2 + \frac{135}{2(938)}\right) = 280MeV$$

2-25. Positronium at rest: $\left(2mc^2\right)^2 = E_i^2 + \left(p_ic\right)^2$

Because $\mathbf{p_i} = 0$, $E_i = 2mc^2 = 2(0.511MeV) = 1.022MeV$

After photon creation; $\left(2mc^2\right)^2 = E_f^2 + \left(p_fc\right)^2$

Because $\mathbf{p_f} = 0$ and energy is conserved, $\left(2mc^2\right)^2 = E_f^2 = \left(1.022MeV\right)^2$ or

$2mc^2 = 1.022MeV$ for the photons.

2-29. $E^2 = \left(pc^2\right)^2 + \left(mc^2\right)^2$ (Equation 2-31)

$$\left(1746MeV\right)^2 = \left(500MeV\right)^2 + \left(mc^2\right)^2$$

$$mc^2 = \left[\left(1746MeV\right)^2 - \left(500MeV\right)^2\right]^{1/2} = 1673MeV \rightarrow m = 1673MeV/c^2$$

$$E = \gamma mc^2 \rightarrow \gamma = 1/\sqrt{1 - u^2/c^2} = E/mc^2$$

$$u/c = \left[1 - \left(mc^2/E\right)^2\right]^{1/2} = \left[1 - \left(1673MeV/1746MeV\right)^2\right]^{1/2} = 0.286 \rightarrow u = 0.286c$$

2-33. Because the clock furthest from Earth (where Earth's gravity is less) runs the faster, answer (c) is correct.

2-37. The speed v of the satellite is:

$$v = 2\pi R / T = 2\pi \left(6.37 \times 10^6 \, m\right) / \left(90 \min \times 60s / \min\right) = 7.42 \times 10^3 \, m/s$$

Special relativistic effect:

After one year the clock in orbit has recorded time $\Delta t = \Delta t / \gamma$, and the clocks differ by:

$$\Delta t - \Delta t' = \Delta t - \Delta t / \gamma = \Delta t \left(1 - 1/\gamma\right) \approx \Delta t \left(v^2 / 2c^2\right), \text{ because } v \ll c. \text{ Thus,}$$

$$\Delta t - \Delta t' = \left(3.16 \times 10^7 s\right)\left(7.412 \times 10^3\right)^2 / \left(2\right)\left(3.00 \times 10^8 \, m\right)^2 = 0.00965s = 9.65 ms$$

Due to special relativity time dilation the orbiting clock is behind the Earth clock by 9.65 ms.

General relativistic effect:

$$\frac{\Delta f}{f_0} = \frac{gh}{c^2} = \frac{\left(9.8 m/s^2\right)\left(3.0 \times 10^5 \, m\right)}{\left(3.0 \times 10^8 \, m/s\right)^2} = 3.27 \times 10^{-11} s/s$$

In one year the orbiting clock gains $\left(3.27 \times 10^{-11} s/s\right)\left(3.16 \times 10^7 s/y\right) = 1.03 ms$.

The net difference due to both effects is a slowing of the orbiting clock by 9.65−1.03 = 8.62 ms.

2-41. (a) The momentum of the ejected fuel is:

$$p = \gamma m u = m u / \sqrt{1 - u^2 / c^2} = 10^3 kg \left(c/2\right) / \sqrt{1 - \left(0.5\right)^2} = 1.73 \times 10^{11} kg \cdot m/s$$

Conservation of momentum requires that this also be the momentum p_s of the spaceship: $p_s = m_s u_s / \sqrt{1 - u_s^2 / c^2} = 1.73 \times 10^{11} kg \cdot m/s$

or, $m_s u_s / \sqrt{1 - u_s^2 / c^2} = \left(1.73 \times 10^{11} kg \cdot m/s\right)^2$

$m_s^2 c_s^2 = \left(1 - u_s^2 / c^2\right)\left(1.73 \times 10^{11} kg \cdot m/s\right)^2 = \left(1.73 \times 10^{11} kg \cdot m/s^2\right) - \left(3.33 \times 10^5 kg^2\right) u_s^2$

$\left(10^6 kg\right)^2 u_s^2 + \left(3.33 \times 10^5 kg^2\right) u_s^2 = \left(1.73 \times 10^{11} kg \cdot m/s\right)^2$

or, $u_s = \left(1.73 \times 10^{11} kg \cdot m/s\right) / 10^6 kg = 1.73 \times 10^5 \, m/s = 5.77 \times 10^{-4} c$

(Problem 2-41 continued)

(b) In classical mechanics, the momentum of the ejected fuel is: $mu = mc/2 = 10^3 c/2$, which must equal the magnitude of the spaceship's momentum $m_s u_s$, so

$$u_s = 10^3 (c/2)/m_s = \frac{10^3 kg \left(3.0 \times 10^8 m/s\right)}{2\left(10^6 kg\right)} = 5.0 \times 10^{-4} c = 1.5 \times 10^5 m/s$$

(c) The initial energy E_i before the fuel was ejected is $E_i = m_s c^2$ in the ship's rest system. Following fuel ejection, the final energy E_f is:

$$E_f = \text{energy of fuel} + \text{energy of ship} = mc^2/\sqrt{1-u^2/c^2} + (m_s - m)c^2/\sqrt{1-u_s^2/c^2}$$

where $u = 0.5c$ and $u_s \ll c$, so $E_f = 1.155 mc^2 + (m_s - m)c^2 = (1.155-1)mc^2 + m_s c^2$

The change in energy ΔE is:

$$\Delta E = E_f - E_i = \left[(0.155)\left(10^3 kg\right)c^2 + \left(10^6 kg\right)c^2\right] - \left[\left(10^6 kg\right)c^2\right]$$

$$\Delta E = \left(155 kg\right)c^2 \text{ or } 155 \ kg = \Delta E/c^2 \text{ of mass has been converted to energy.}$$

2-45. The minimum energy photon needed to create an $e^- - e^+$ pair is $E_p = 1.022 \ MeV$ (see Example 2-13). At minimum energy, the pair is created at rest, i.e., with no momentum. However, the photon's momentum must be $p = E/c = 1.022 MeV/c$ at minimum. Thus, momentum conservation is violated unless there is an additional mass "nearby" to absorb recoil momentum.

2-49. (a) $p_i = 0 = E/c - Mv$ or $v = E/Mc$

(b) The box moves a distance $\Delta x = v\Delta t$, where $\Delta t = L/c$,

so $\Delta x = \left(E/Mc\right)\left(L/c\right) = EL/Mc^2$

(c) Let the center of the box be at $x = 0$. Radiation of mass m is emitted from the left end of the box (e.g.) and the center of mass is at:

$$x_{CM} = \frac{M(0) + m(L/2)}{M + m} = \frac{mL}{2(M + m)}$$

(Problem 2-49 continued)

When the radiation is absorbed at the other end the center of mass is at:

$$x_{CM} = \frac{M\left(EL/Mc^2\right) + m\left(L/2 - EL/Mc^2\right)}{M + m}$$

Equating the two values of x_{CM} (if CM is not to move) yields:

$$m = \left(E/c^2\right)/\left(1 - E/Mc^2\right)$$

Because $E \ll Mc^2$, then $m \approx E/c^2$ and the radiation has this mass.

2-53. (a) $F_x' = \dfrac{dp_x}{dt} = \dfrac{d\left(\gamma m u_x'\right)}{dt}$ $F_x = ma_x$ because $u_x = 0$

$$F_x' = \gamma m\left(du_x'/dt\right) + m u_x'\, d\left[\left(1 - u_x'^2/c^2\right)^{-1/2}\right]/dt$$

$$F_x' = \frac{ma_x'}{\left(1 - u_x'^2/c^2\right)^{1/2}} + \frac{m\left(u_x'^2/c^2\right)a_x'}{\left(1 - u_x'^2/c^2\right)^{3/2}}$$

$$F_x' = \frac{ma_x'\left(1 - u_x'^2/c^2\right) + m\left(u_x'^2/c^2\right)a_x'}{\left(1 - u_x'^2/c^2\right)^{3/2}}$$

$$F_x' = \gamma^3 m a_x'$$

Because $u_x = 0$, note from Equation 2-1 that $a_x' = a_x/\gamma^3$.

Therefore, $F_x' = \gamma^3 m a_x/\gamma^3 = ma_x = F_x$

(b) $F_y' = \dfrac{dp_y'}{dt'} = \dfrac{d\left(\gamma m u_y'\right)}{dt'}$ $F_y = ma_y$ because $u_y = u_x = 0$

$F_y' = \gamma m a_y'$ because S' moves in the $+x$ direction and the instantaneous transverse impulse (small) changes only the direction of $\mathbf{v}$. From the result of Problem 2-5 (inverse form) with $u_y = u_x = 0$, $a_y' = a_y/\gamma^2$

Therefore, $F_y' = \gamma m a_y' = \gamma m a_y/\gamma^2 = ma_y/\gamma = F_y/\gamma$

Chapter 3 – Quantization of Charge, Light, and Energy

3-1.   The radius of curvature is given by Equation 3-2.

$$R = \frac{mu}{qB} = m\left[\frac{2.5\times10^6\,m/s}{\left(1.60\times10^{-19}C\right)\left(0.40T\right)}\right] = m\left(3.91\times10^{25}\,m/s\cdot C\cdot T\right)$$

Substituting particle masses from Appendices A and D:

$$R\,(\text{protron}) = \left(1.67\times10^{-27}\,kg\right)\left(3.91\times10^{25}\,m/s\cdot C\cdot T\right) = 6.5\times10^{-2}\,m$$

$$R\,(\text{electron}) = \left(9.11\times10^{-31}\,kg\right)\left(3.91\times10^{25}\,m/s\cdot C\cdot T\right) = 3.6\times10^{-5}\,m$$

$$R\,(\text{deuteron}) = \left(3.34\times10^{-27}\,kg\right)\left(3.91\times10^{25}\,m/s\cdot C\cdot T\right) = 0.13\,m$$

$$R\,(H_2) = \left(3.35\times10^{-27}\,kg\right)\left(3.91\times10^{25}\,m/s\cdot C\cdot T\right) = 0.13\,m$$

$$R\,(\text{helium}) = \left(6.64\times10^{-27}\,kg\right)\left(3.91\times10^{25}\,m/s\cdot C\cdot T\right) = 0.26\,m$$

3-5.   (a)

$$R = \frac{mu}{qB} = \frac{\left[\left(2E_k/e\right)\left(e/m\right)\right]^{1/2}}{\left(e/m\right)\left(B\right)}$$

$$= \frac{1}{B}\sqrt{\frac{2E_k/e}{e/m}} = \frac{1}{0.325T}\left[\frac{\left(2\right)\left(4.5\times10^4\,eV/e\right)}{1.76\times10^{11}\,kg}\right]^{1/2} = 2.2\times10^{-3}\,m = 2.2\,mm$$

(b)

frequency $\quad f = \dfrac{u}{2\pi R} = \dfrac{\sqrt{\left(2E_k/e\right)\left(e/m\right)}}{2\pi R}$

$$\vdots \quad = \frac{\left[\left(2\right)\left(4.5\times10^4\,eV/e\right)\left(1.76\times10^{11}\,C/kg\right)\right]^{1/2}}{2\pi\left(2.2\times10^{-3}\,m\right)} = 9.1\times10^9\,Hz$$

period $\quad T = 1/f = 1.1\times10^{-10}\,s$

3-9.   For the rise time to equal the field-free fall time, the net upward force must equal the weight. $q\mathrm{E} - mg = mg$ $\therefore$ $\mathrm{E} = 2mg/q$.

3-13.   Equation 3-4: $R = \sigma T^4$. Equation 3-6: $R = \dfrac{1}{4}cU$.

From Example 3-4: $U = \left(8\pi^5 k^4 T^4\right)/\left(15h^3 c^2\right)$

$$\sigma = \frac{R}{T^4} = \frac{(1/4)cU}{T^4} = \frac{1}{4}c\left(8\pi^5 k^4 T^4\right)/\left(15h^3 c^2 T^4\right)$$

$$= \frac{2\pi^5\left(1.38\times10^{-23}\,J/K\right)^4}{15\left(6.63\times10^{-34}\,J\cdot s\right)^3\left(3.00\times10^8\,m/s\right)^2} = 5.67\times10^{-8}\,W/m^2 K^4$$

3-17.   Equation 3-4: $R_1 = \sigma T_1^4$ $R_2 = \sigma T_2^4 = \sigma\left(2T_1\right)^4 = 16\sigma T_1^4 = 16R_1$

3-21.   Equation 3-4: $R = \sigma T^4$

$$P_{abs} = \left(1.36\times10^3\,W/m^2\right)\left(\pi R_E^2 m^2\right) \text{ where } R_E = \text{radius of Earth}$$

$$P_{emit} = \left(RW/m^2\right)\left(4\pi R_E^2\right) = \left(1.36\times10^3\,W/m^2\right)\left(\pi R_E^2 m^2\right)$$

$$R = \left(1.36\times10^3\,W/m^2\right)\left(\frac{\pi R_E^2}{4\pi R_E^2}\right) = \frac{1.36\times10^3}{4}\frac{W}{m^2} = \sigma T^4$$

$$T^4 = \frac{1.36\times10^3\,W/m^2}{4\left(5.67\times10^{-8}\,W/m^2\cdot K^4\right)}\quad \therefore\quad T = 278.3K = 5.3°C$$

3-25.   (a) $hf = hc/\lambda = 0.47eV$.

$$\lambda_{max} = \frac{hc}{4.87eV} = \frac{\left(4.14\times10^{-15}\,eV\cdot s\right)\left(3.00\times10^8\,m/s\right)}{4.87eV} = 2.55\times10^{-7}\,m = 255nm$$

(Problem 3-25 continued)

(b) It is the fraction of the total solar power with wavelengths less than $255nm$, i.e., the area under the Planck curve (Figure 3-6) up to $255nm$ divided by the total area. The latter is: $R = \sigma T^4 = \left(5.67 \times 10^{-8} W/m^2 \cdot K^4\right)\left(5800K\right)^4 = 6.42 \times 10^7 W/m^2$.

Approximating the former with $u(\lambda)\Delta\lambda$ with $\lambda = 127nm$ and $\Delta\lambda = 255nm$:

$$\left[u\left(127nm\right)\right]\left(255nm\right) = \left[\frac{8\pi hc\left(127 \times 10^{-9}m\right)^{-5}}{e^{hc/kT\left(127 \times 10^{-9}\right)} - 1}\right]\left(255 \times 10^9 m\right) = 1.23 \times 10^{-4} J/m^3$$

$$R\left(0 - 255nm\right) = \frac{c}{4}\left(1.23 \times 10^{-4} J/m^3\right) \quad \rightarrow \quad \frac{R\left(0 - 255nm\right)}{R}$$

$$= \frac{\left(3.00 \times 10^8 m/s\right)\left(1.23 \times 10^4 J/m^3\right)}{\left(4\right)\left(6.42 \times 10^7 W/m^2\right)} \qquad \text{fraction} = 1.4 \times 10^{-4}$$

3-29. (a) $E = hf = hc/\lambda \quad \Rightarrow \quad \lambda = hc/E$

For $E = 4.26eV$: $\quad \lambda = \left(1240eV \cdot nm\right)/\left(4.26eV\right) = 291nm$

and since $f = c/\lambda$, $\quad f = \left(3.00 \times 10^8 m/s\right)/\left(291nm\right) = 1.03 \times 10^{15} s^{-1}$

(b) This photon is in the ultraviolet region of the electromagnetic spectrum.

3-33. Equation 3-25: $\lambda_2 - \lambda_1 = \Delta\lambda = \frac{h}{mc}\left(1 - \cos\theta\right)$

$$\Delta\lambda = \frac{\left(6.63 \times 10^{-34} J \bullet s\right)\left(1 - \cos 135°\right)}{\left(9.11 \times 10^{-31} kg\right)\left(3.00 \times 10^8 m/s\right)} = 4.14 \times 10^{-12} m = 4.14 \times 10^{-3} nm$$

$$\frac{\Delta\lambda}{\lambda_1} \times 100 = \frac{4.14 \times 10^{-3} nm}{0.0711 nm} \times 100 = 5.8\%$$

3-37. $\Delta\lambda = \lambda_2 - \lambda_1 = \Delta\lambda = \frac{h}{mc}\left(1 - \cos\theta\right) = 0.01\lambda_1$ Equation 3-25

$$\lambda_1 = \left(100\right)\frac{h}{mc}\left(1 - \cos\theta\right) = \left(100\right)\left(0.00243nm\right)\left(1 - \cos 90°\right) = 0.243nm$$

3-41.   (a) Compton wavelength $= \dfrac{h}{mc}$

electron: $\dfrac{h}{mc} = \dfrac{6.63 \times 10^{-34} \, J \bullet s}{\left(9.11 \times 10^{-31} kg\right)\left(3.00 \times 10^{8} \, m/s\right)} = 2.43 \times 10^{-12} \, m = 0.00243 nm$

proton: $\dfrac{h}{mc} = \dfrac{6.63 \times 10^{-34} \, J \bullet s}{\left(1.67 \times 10^{-27} kg\right)\left(3.00 \times 10^{8} \, m/s\right)} = 1.32 \times 10^{-15} \, m = 1.32 \, fm$

(b) $E = \dfrac{hc}{\lambda}$

(i) electron: $E = \dfrac{1240 eV \bullet nm}{0.00243 nm} = 5.10 \times 10^{5} \, eV = 0.510 MeV$

(ii) proton: $E = \dfrac{1240 eV \bullet nm}{1.32 \times 10^{-6} \, nm} = 9.39 \times 10^{8} \, eV = 939 MeV$

3-45.   Calculate $1/\lambda$ to be used in the graph.

$1/\lambda \ (10^{6}/m)$	5.0	3.3	2.5	2.0	1.7
V_0 (V)	4.20	2.06	1.05	0.41	0.03

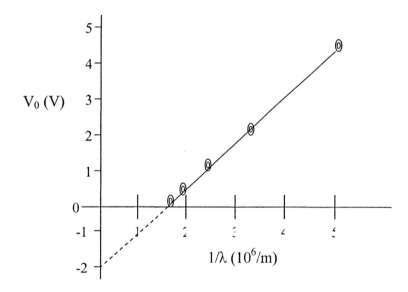

(a) The intercept on the vertical axis is the work function ϕ. $\phi = 2.08 eV$.

(Problem 3-45 continued)

(b) The intercept on the horizontal axis corresponds to the threshold frequency.

$$\frac{1}{\lambda_t} = 1.65 \times 10^6 \, / \, m$$

$$f_t = \frac{c}{\lambda_t} = \left(3.00 \times 10^8 \, m/s\right)\left(1.65 \times 10^6 \, /\, m\right) = 4.95 \times 10^{14} \, Hz$$

(c) The slope of the graph is h/e. Using the vertical intercept and the largest experimental point,

$$\frac{h}{e} = \frac{1}{c} \frac{\Delta V_0}{\Delta(1/\lambda)} = \frac{4.20V - (-2.08V)}{\left(3.00 \times 10^8 \, m/s\right)\left(5.0 \times 10^6 \, /\, m - 0\right)} = 4.19 \times 10^{-15} \, eV \, / \, Hz$$

3-49.   Conservation of energy: $E_1 + mc^2 = E_2 + E_k + mc^2$ $\therefore$ $E_k = E_1 - E_2 = hf_1 - hf_2$

From Compton's equation, we have: $\lambda_2 - \lambda_1 = \dfrac{h}{mc}(1 - \cos\theta)$,

Thus, $\dfrac{1}{f_2} - \dfrac{1}{f_1} = \dfrac{h}{mc^2}(1 - \cos\theta)$

$$\frac{1}{f_2} = \frac{1}{f_1} + \frac{h}{mc^2}(1 - \cos\theta) \quad \therefore \quad f_2 = \frac{f_1 mc^2}{mc^2 + hf_1(1 - \cos\theta)}$$

Substituting this expression for f_2 into the expression for E_k (and dropping the subscript on f_1):

$$E_k = hf - \frac{hfmc^2}{mc^2 + hf(1 - \cos\theta)} = \frac{hfmc^2 + (hf)^2(1 - \cos\theta) - hfmc^2}{mc^2 + hf(1 - \cos\theta)} = \frac{hf}{1 + \dfrac{mc^2}{\left[hf(1 - \cos\theta)\right]}}$$

E_k has its maximum value when the photon energy change is maximum, i.e., when $\theta = \pi$

so $\cos\theta = -1$. Then $E_k = \dfrac{hf}{1 + \dfrac{mc^2}{2hf}}$

3-53. (a) Equation 3-18: $u(\lambda) = \dfrac{8\pi hc\lambda^{-5}}{e^{hc/\lambda kT}-1}$ Letting $C = 8\pi hc$ and $a = hc/kT$

gives $u(\lambda) = \dfrac{C\lambda^{-5}}{e^{a/\lambda}-1}$

(b)

$$\frac{du}{d\lambda} = \frac{d}{d\lambda}\left[\frac{C\lambda^{-5}}{e^{a/\lambda}-1}\right] = C\left[\frac{\lambda^{-5}(-1)e^{a/\lambda}(-a\lambda^{-2})}{\left(e^{a/\lambda}-1\right)^2} - \frac{5\lambda^{-6}}{e^{a/\lambda}-1}\right]$$

$$= \frac{C\lambda^{-6}}{\left(e^{a/\lambda}-1\right)^2}\left[\frac{a}{\lambda}e^{a/\lambda} - 5\left(e^{a/\lambda}-1\right)\right] = \frac{C\lambda^{-6}e^{a/\lambda}}{\left(e^{a/\lambda}-1\right)^2}\left[\frac{a}{\lambda} - 5\left(1-e^{a/\lambda}\right)\right] = 0$$

The maximum corresponds to the vanishing of the quantity in brackets.
Thus, $5\lambda\left(1-e^{-a/\lambda}\right) = a$

(c) This equation is most efficiently solved by trial and error; i.e., guess at a value for

a/λ in the expression $5\lambda\left(1-e^{-a/\lambda}\right) = a$, solve for a better value of a/λ; substitute

the new value to get an even better value, and so on. Repeat the process until the

calculated value no longer changes. One succession of values is 5, 4.966310,

4.965156, 4.965116, 4.965114, 4.965114. Further iterations repeat the same value (to

seven digits), so we have $\dfrac{a}{\lambda_m} = 4.965114 = \dfrac{hc}{\lambda_m kT}$

(d) $\lambda_m T = \dfrac{hc}{(4.965114)k} = \dfrac{\left(6.63\times10^{-34}\,J\bullet s\right)\left(3.00\times10^8\,m/s\right)}{(4.965114)\left(1.38\times10^{-23}\,J/K\right)}$

Therefore, $\lambda_m T = 2.898\times10^{-3}\,m\bullet K$ Equation 3-5

3-57. (a) $E_k = 50\,keV$ and $\lambda_2 = \lambda_1 + 0.095\,nm$

$$\frac{hc}{\lambda_1} + \frac{hc}{\lambda_2} = 5.0\times10^4\,eV \quad \therefore \quad \frac{1}{\lambda_1} + \frac{1}{\lambda_1 + 0.095} = \frac{5.0\times10^4\,eV}{hc}$$

$$\therefore \quad \frac{2\lambda_1 + 0.095}{\lambda_1^2 + 0.095\lambda_1} = \frac{5.0\times10^4\,eV}{hc}$$

$$\lambda_1^2 + \left(0.095\,nm - \frac{2hc}{5\times10^4\,eV}\right)\lambda_1 - \frac{(0.095\,nm)hc}{5\times10^4\,eV} = 0$$

(Problem 3-57 continued)

$$\therefore \quad \lambda_1^2 + 0.04541\lambda_1 - 2.36 \times 10^{-3} = 0$$

Applying the quadratic formula,

$$\lambda_1 = \frac{-0.04541 \pm \left[(0.04541)^2 + 4(2.36 \times 10^{-3})\right]^{1/2}}{2}$$

$$\lambda_1 = 0.03092 nm \text{ and } \lambda_2 = 0.1259 nm$$

(b) $E_1 = \dfrac{hc}{\lambda_1} = \dfrac{1240 eV \cdot nm}{0.03092 nm} = 40.1 keV \quad \rightarrow \quad E_{electron} = 9.90 keV$

Chapter 4 – The Nuclear Atom

4-1. $\dfrac{1}{\lambda_{mn}} = R\left(\dfrac{1}{m^2} - \dfrac{1}{n^2}\right)$ where $R = 1.097 \times 10^7 \, m^{-1}$ (Equation 4-2)

The Lyman series ends on $m = 1$, the Balmer series on $m = 2$, and the Paschen series on $m = 3$. The series limits all have $n = \infty$, so $\dfrac{1}{n} = 0$.

$$\frac{1}{\lambda_L} = R\left(\frac{1}{1^2}\right) = 1.097 \times 10^7 \, m^{-1}$$

$$\lambda_L(\text{limit}) = 1.097 \times 10^7 \, m^{-1} = 91.16 \times 10^{-9} \, m = 91.16 \, nm$$

$$\frac{1}{\lambda_B} = R\left(\frac{1}{2^2}\right) = 1.097 \times 10^7 \, m^{-1} / 4$$

$$\lambda_B(\text{limit}) = 4/1.097 \times 10^7 \, m^{-1} = 3.646 \times 10^{-7} \, m = 364.6 \, nm$$

$$\frac{1}{\lambda_P} = R\left(\frac{1}{3^2}\right) = 1.097 \times 10^7 \, m^{-1} / 9$$

$$\lambda_P(\text{limit}) = 9/1.097 \times 10^7 \, m^{-1} = 8.204 \times 10^{-7} \, m = 820.4 \, nm$$

4-5. None of these lines are in the Paschen series, whose limit is 820.4 nm (see Problem 4-1) and whose first line is given by $\dfrac{1}{\lambda_{34}} = R\left(\dfrac{1}{3^2} - \dfrac{1}{4^2}\right) \rightarrow \lambda_{34} = 1875 \, nm$. Also, none are in the Brackett series, whose longest wavelength line is 4052 nm (see Problem 4-4). The Pfund series has $m = 5$. Its first three (i.e., longest wavelength) lines have $n = 6$, 7, and 8.

$$\frac{1}{\lambda_{56}} = 1.097 \times 10^7 \, m^{-1}\left(\frac{1}{5^2} - \frac{1}{5^2}\right) = 1.341 \times 10^5 \, m^{-1}$$

$$\lambda_{56} = \frac{1}{1.341 \times 10^5 \, m^{-1}} = 7.458 \times 10^{-6} \, m = 7458 \, nm. \text{ Similarly,}$$

$$\lambda_{57} = \frac{1}{2.155 \times 10^5 \, m^{-1}} = 4.653 \times 10^{-6} \, m = 4653 \, nm$$

(Problem 4-5 continued)

$$\lambda_{58} = \frac{1}{2.674 \times 10^5 \, m^{-1}} = 3.740 \times 10^{-6} \, m = 3740 \, nm$$

Thus, the line at 4103 *nm* is not a hydrogen spectral line.

4-9. $\quad r_d = \dfrac{kq_\alpha Q}{(1/2) m_\alpha v^2} = \dfrac{ke^2 \cdot 2 \cdot 79}{E_{k\alpha}}$ (Equation 4-11)

For $E_{k\alpha} = 5.0 MeV$: $\quad r_d = \dfrac{(1.44 MeV \cdot fm)(2)(79)}{5.0 MeV} = 45.5 \, fm$

For $E_{k\alpha} = 7.7 MeV$: $\quad r_d = 29.5 \, fm$

For $E_{k\alpha} = 12 MeV$: $\quad r_d = 19.0 \, fm$

4-13. (a) $\quad r_n = \dfrac{n^2 a_0}{Z}$ (Equation 4-18)

$$r_6 = \frac{6^2 (0.053 nm)}{1} = 1.91 nm$$

(b) $\quad r_6 \left(He^+ \right) = \dfrac{6^2 (0.053 nm)}{2} = 0.95 nm$

4-17. $\quad f_{rev} = \dfrac{mk^2 Z^2 e^4}{2\pi\hbar^3 n^3}$ (Equation 4-29)

$$= \frac{mc^2 Z^2 \left(ke^2\right)^2}{2\pi\hbar n^3 (\hbar c)^2} = \frac{cZ^2}{(h/mc)n^3}\left(\frac{ke^2}{\hbar c}\right)^2 = \frac{cZ^2 \alpha^2}{\lambda_c n^3}$$

$$= \frac{\left(3.00 \times 10^8 \, m/s\right)(1)^2}{\left(0.00243 \times 10^{-9} \, m\right)(2)^3}\left(\frac{1}{137}\right)^2 = 8.22 \times 10^{14} \, Hz$$

$$N = f_{rev} t = \left(8.22 \times 10^{14} \, Hz\right)\left(10^{-8} \, s\right) = 8.22 \times 10^6 \text{ revolutions}$$

4-21.

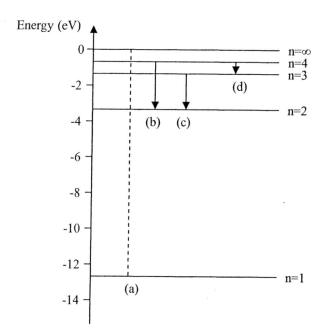

(a) Lyman limit, (b) H_β line, (c) H_α line, (d) longest wavelength line of Paschen series

4-25. (a) The radii of the Bohr orbits are given by (see Equation 4-18)

$r = n^2 a_0 / Z$ where $a_0 = 0.0529nm$ and $Z = 1$ for hydrogen.

For $n = 600$, $r = (600)^2 (0.0529nm) = 1.90 \times 10^4 nm = 19.0 \mu m$

This is about the size of a tiny grain of sand.

(b) The electron's speed in a Bohr orbit is given by

$v^2 = ke^2 / mr$ with $Z = 1$

Substituting r for the $n = 600$ orbit from (a), then taking the square root,

$v^2 = (8.99 \times 10^9 N \bullet m^2)(1.609 \times 10^{-19} C)^2 / (9.11 \times 10^{-31} kg)(19.0 \times 10^{-6} m)$

$v^2 = 1.33 \times 10^7 m^2 / s^2 \quad \rightarrow \quad v = 3.65 \times 10^3 m/s$

For comparison, in the $n = 1$ orbit, v is about $2 \times 10^6 m/s$

4-29. $r_n = \dfrac{n^2 a_0}{Z}$ (Equation 4-18)

The $n = 1$ electrons "see" a nuclear charge of approximately $Z - 1$, or 78 for Au.

$r_1 = 0.0529nm / 78 = 6.8 \times 10^{-4} nm (10^{-9} m/nm)(10^{15} fm/m) = 680 fm$, or about 100 times

the radius of the Au nucleus.

4-33.

Element	Al	Ar	Sc	Fe	Ge	Kr	Zr	Ba
Z	13	18	21	26	32	36	40	56
E (keV)	1.56	3.19	4.46	7.06	10.98	14.10	17.66	36.35
$f^{1/2}\left(10^8\,Hz^{1/2}\right)$	6.14	8.77	10.37	13.05	16.28	18.45	20.64	29.62

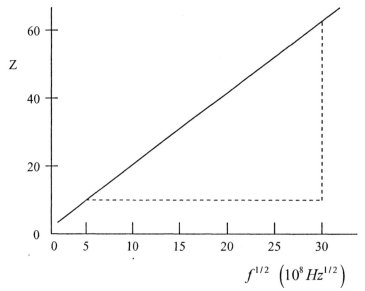

$$\text{slope} = \frac{58-10}{(30-4.8)\times10^8} = 1.90\times10^{-8}\,Hz^{-1/2}$$

$$\text{slope (Figure 4-19)} = \frac{30-13}{(5-7)\times10^8} = 2.13\times10^{-8}\,Hz^{-1/2}$$

The two values are in good agreement.

4-37. Using the results from Problem 4-24, the energy of the positronium Lyman α line is

$$\Delta E = E_2 - E_1 = -1.701eV - (-6.804eV) = 5.10eV.$$ The first Franck-Hertz current decrease would occur at $5.10V$, the second at $10.2V$.

4-41. (a)

$$i = qf_{rev} = e\frac{Z^2mk^2e^4}{2\pi\hbar^3n^3} \quad \text{(from Equation 4-28)}$$

$$= e\frac{mc^2\left(ke^2\right)^2(1)^2}{2\pi\hbar(\hbar c)^2(1)^3} = \frac{ec}{(h/mc)}\left(\frac{ke^2}{\hbar c}\right)^2 = \frac{ec\alpha^2}{\lambda_c}$$

$$= \frac{\left(1.602\times10^{-19}C\right)\left(3.00\times10^{17}nm/s\right)}{0.00243nm}\left(\frac{1}{137}\right)^2 = 1.054\times10^{-3}A$$

(b)

$$\mu = iA = i\pi a_0^2 = \left(\frac{emk^2e^4}{2\pi\hbar^3}\right)\pi\left(\frac{\hbar^2}{mke^2}\right) = \frac{e\hbar}{2m}$$

$$= \frac{\left(1.602\times10^{-19}C\right)\left(1.055\times10^{-34}J\cdot s\right)}{2\left(9.11\times10^{-31}kg\right)} = 9.28\times10^{-24}A\cdot m^2$$

or

$$= \left(1.054\times10^{-3}A\right)\pi\left(0.529\times10^{-10}m\right)^2 = 9.27\times10^{-24}A\cdot m^2$$

4-45. (a) $E_n = -E_0Z^2/n^2$ (Equation 4-20)

For Li^{++}, $Z = 3$ and $E_n = -13.6eV(9)/n^2 = -122.4/n^2 eV$

The first three Li^{++} levels that have the same (nearly) energy as H are:

$n = 3$, $E_3 = -13.6eV$ $\quad$ $n = 6$, $E_6 = -3.4eV$ $\quad$ $n = 9$, $E_9 = -1.51eV$

Lyman α corresponds to the $n = 6 \to n = 3$ Li^{++} transitions. Lyman β corresponds to the $n = 9 \to n = 3$ Li^{++} transition.

(b) $R(H) = R_\infty\left(1/\left(1+0.511MeV/938.8MeV\right)\right) = 1.096776\times10^7 m^{-1}$

$R(Li) = R_\infty\left(1/\left(1+0.511MeV/6535MeV\right)\right) = 1.097287\times10^7 m^{-1}$

For Lyman α:

$$\frac{1}{\lambda} = R(H)\left(1-\frac{1}{2^2}\right) = 1.096776\times10^7 m^{-1}\left(10^{-9}m/nm\right)(3/4) \to 121.568nm$$

For Li^{++} equivalent:

$$\frac{1}{\lambda} = R(Li)\left(\frac{1}{3^2}-\frac{1}{6^2}\right)Z^2 = 1.097287\times10^7 m^{-1}\left(10^{-9}m/nm\right)\left(\frac{1}{9}-\frac{1}{36}\right)(3)^2$$

$$\lambda = 121.512nm \quad \Delta\lambda = 0.056nm$$

4-49. (a) $b = R\sin\beta = R\sin\left(\dfrac{180° - \theta}{2}\right) = R\cos\dfrac{\theta}{2}$

(b) Scattering through an angle larger than θ corresponds to an impact parameter smaller than b. Thus, the shot must hit within a circle of radius b and area πb^2. The rate at which this occurs is $I_0 \pi b^2 = I_0 R^2 \cos^2\dfrac{\theta}{2}$

(c) $\sigma = \pi b_0^2 = \pi\left(R\cos\dfrac{\theta}{2}\right)^2 = \pi R^2$

(d) An α particle with an arbitrarily large impact parameter still feels a force and is scattered.

4-53. $\dfrac{kZe^2}{r} = \dfrac{mv^2}{r} \rightarrow \dfrac{kZe^2}{r^2} = \dfrac{(\gamma mv)^2}{mr}$ (from Equation 4-12)

$\gamma v = \left(\dfrac{kZe^2}{mr}\right)^{1/2} = \dfrac{v}{\sqrt{1-\beta^2}}$

$\dfrac{c^2\beta^2}{1-\beta^2} = \left(\dfrac{kZe^2}{mr}\right)$ Therefore, $\beta^2\left[c^2 + \left(\dfrac{kZe^2}{mr}\right)\right] = \left(\dfrac{kZe^2}{mr}\right)$

$\beta^2 \approx \dfrac{1}{c^2}\left(\dfrac{kZe^2}{ma_o}\right) \rightarrow \beta = 0.0075Z^{1/2} \rightarrow v = 0.0075cZ^{1/2} = 2.25\times10^6\,m/s\times Z^{1/2}$

$E = KE - kZe^2/r = mc^2(\gamma-1) - \dfrac{kZe^2}{r} = mc^2\left[\dfrac{1}{\sqrt{1-\beta^2}} - 1\right] - \dfrac{kZe^2}{r}$

And substituting $\beta = 0.0075$ and $r = a_o$

$E = 511\times10^3 eV\left[\dfrac{1}{\sqrt{1-(0.0075)^2}} - 1\right] - 28.8Z\ eV$

$= 14.4eV - 28.8Z\ eV = -14.4Z\ eV$

4-57. Refer to Figure 4-16. All possible transitions starting at $n = 5$ occur.

$n = 5$ to $n = 4, 3, 2, 1$

$n = 4$ to $n = 3, 2, 1$

$n = 3$ to $n = 2, 1$

$n = 2$ to $n = 1$

Thus, there are 10 different photon energies emitted.

n_i	n_f	fraction	no. of photons
5	4	$1/4$	125
5	3	$1/4$	125
5	2	$1/4$	125
5	1	$1/4$	125
4	3	$1/4 \times 1/3$	42
4	2	$1/4 \times 1/3$	42
4	1	$1/4 \times 1/3$	42
3	2	$1/2\left[1/4 + 1/4(1/3)\right]$	83
3	1	$1/2\left[1/4 + 1/4(1/3)\right]$	83
2	1	$\left[\left(1/2(1/4 + 1/4)(1/3)\right) + 1/4(1/3) + 1/4\right]$	250

Total = 1,042

Note that the number of electrons arriving at the $n = 1$ level (125+42+83+250) is 500, as it should be.

Chapter 5 – The Wavelike Properties of Particles

5-1.   (a) $\lambda = \dfrac{h}{p} = \dfrac{h}{mv} = \dfrac{\left(6.63\times10^{-34}\,J\bullet s\right)\left(3.16\times10^{7}\,s/y\right)}{\left(10^{-3}\,kg\right)\left(1m/y\right)} = 2.1\times10^{-23}\,m$

     (b) $v = \dfrac{h}{m\lambda} = \dfrac{6.63\times10^{-34}\,J\bullet s}{\left(10^{-3}\,kg\right)\left(10^{-2}\,m\right)} = 6.6\times10^{-29}\,m/s = 2.1\times10^{-21}\,m/y$

5-5. $\lambda = h/p = h/\sqrt{2mE_k} = hc/\left[2mc^2\left(1.5kT\right)\right]^{1/2}$ (from Equation 5-2)

Mass of N_2 molecule =
$$2\times14.0031u\left(931.5MeV/uc^2\right) = 2.609\times10^{4}\,MeV/c^2 = 2.609\times10^{10}\,eV/c^2$$

$$\lambda = \dfrac{1240eV\bullet nm}{\left[(2)\left(2.609\times10^{10}\,eV\right)(1.5)\left(8.617\times10^{-5}\,eV/K\right)(300K)\right]^{1/2}} = 0.0276nm$$

5-9. $E_k = mc^2\left(\gamma-1\right)$ $p = \gamma mu$

    (a) $E_k = 2GeV$ $mc^2 = 0.938GeV$

      $\gamma-1 = E_k/mc^2 = 2GeV/0.938GeV = 2.132$ Thus, $\gamma = 3.132$

      Because, $\gamma = 1/\sqrt{1-\left(u/c\right)^2}$ where $u/c = 0.948$

$$\lambda = \dfrac{h}{p} = \dfrac{h}{\gamma mc\left(u/c\right)} = \dfrac{h}{\gamma mc^2\left(u/c\right)}$$

$$= \dfrac{1240eV\bullet nm}{(3.132)\left(938\times10^{6}\,eV\right)(0.948)} = 4.45\times10^{-7}\,nm = 0.445\,fm$$

    (b) $E_k = 200GeV$

      $\gamma-1 = E_k/mc^2 = 200GeV/0.938GeV = 213$ Thus, $\gamma = 214$ and $u/c = 0.9999$

$$\lambda = \dfrac{1240eV\bullet nm}{(214)(938MeV)(0.9999)} = 6.18\times10^{-3}\,fm$$

5-13.

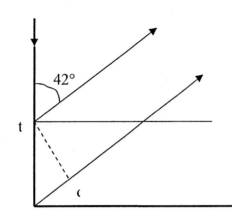

$$d = t\cos 42°$$

$$n\lambda = t + d = t\left(1 + \cos 42°\right) = 0.30nm\left(1 + \cos 42°\right)$$

For the first maximum $n = 1$, so $\lambda = 0.523nm$

$$\lambda = \frac{h}{p} = \frac{h}{\sqrt{2mE_k}} \quad \rightarrow \quad E_k = \frac{h^2}{2m\lambda^2} = \frac{\left(hc\right)^2}{2mc^2\lambda^2}$$

$$E_k = \frac{\left(1240eV\cdot nm\right)^2}{2\left(939 \times 10^6 eV\right)\left(0.523nm\right)} = 3.0 \times 10^{-3} eV$$

5-17. (a) $y = y_1 + y_2$

$$= 0.002m\,\cos\left(8.0x/m - 400t/s\right) + 0.002m\,\cos\left(7.6x/m - 380t/s\right)$$

$$= 2\left(0.002m\right)\cos\left[\frac{1}{2}\left(8.0x/m - 7.6x/m\right) - \frac{1}{2}\left(400t/s - 380t/s\right)\right]$$

$$\times \cos\left[\frac{1}{2}\left(8.0x/m + 7.6x/m\right) - \frac{1}{2}\left(400t/s + 380t/s\right)\right]$$

$$= 0.004m\,\cos\left(0.2x/m - 10t/s\right) \times \cos\left(7.8x/m - 390t/s\right)$$

(b) $v = \dfrac{\overline{\omega}}{\overline{k}} = \dfrac{390/s}{7.8/m} = 50m/s$

(c) $v_s = \dfrac{\Delta\omega}{\Delta k} = \dfrac{20/s}{0.4/m} = 50m/s$

(d) Successive zeros of the envelope requires that $0.2\Delta x/m = \pi$, thus

$$\Delta x = \frac{\pi}{0.2} = 5\pi m \text{ with } \Delta k = k_1 - k_2 = 0.4m^{-1} \text{ and } \Delta x = \frac{2\pi}{\Delta k} = 5\pi m.$$

5-21. $\Delta\omega\,\Delta t \approx 1 \quad \rightarrow \quad \left(2\pi\Delta f\right)\Delta t = 1$ Thus, $\Delta t \approx 1/\left(2\pi \times 5000\right) = 3.2 \times 10^{-5} s$

5-25. (a) At $x = 0$: $Pdx = |\psi(0,0)|^2 \, dx = |Ae^0|^2 \, dx = A^2 dx$

(b) At $x = \sigma$: $Pdx = |Ae^{-\sigma^2/4\sigma^2}|^2 \, dx = |Ae^{-1/4}|^2 \, dx = 0.61A^2 dx$

(c) At $x = 2\sigma$: $Pdx = |Ae^{-4\sigma^2/4\sigma^2}|^2 \, dx = |Ae^{-1}|^2 \, dx = 0.14A^2 dx$

(d) The electron will most likely be found at $x = 0$, where Pdx is largest.

5-29. $\Delta E \Delta t \approx \hbar \quad \rightarrow \quad \Delta E \approx \hbar / \Delta t = \dfrac{6.58 \times 10^{-16}\, eV \cdot s}{3.823d\left(8.64 \times 10^4\, s/d\right)} \approx 1.99 \times 10^{-21}\, eV$

The energy uncertainty of the excited state is ΔE, so the α energy can be no sharper than ΔE.

5-33. (a) For ^{48}Ti:

$$\Delta E\left(\text{upper state}\right) = \frac{\hbar}{\Delta t} = \frac{1.055 \times 10^{-34}\, J \cdot s}{1.4 \times 10^{-14}\, s \left(1.60 \times 10^{-13}\, J/MeV\right)} \approx 4.71 \times 10^{-10}\, MeV$$

$$\Delta E\left(\text{lower state}\right) = \frac{\hbar}{\Delta t} = \frac{1.055 \times 10^{-34}\, J \cdot s}{3.0 \times 10^{-12}\, s \left(1.60 \times 10^{-13}\, J/MeV\right)} \approx 2.20 \times 10^{-10}\, MeV$$

$$\Delta E\left(\text{total}\right) = \Delta E_U + \Delta E_L = 6.91 \times 10^{-10}\, MeV$$

$$\frac{\Delta E_T}{E} = \frac{6.91 \times 10^{-10}\, MeV}{1.312 MeV} = 5.3 \times 10^{-10}$$

(b) For Hα: $\Delta E_U \approx \dfrac{1.055 \times 10^{-34}\, J \cdot s}{10^{-8}\, s \left(1.60 \times 10^{-19}\, J/eV\right)} \approx 6.59 \times 10^{-8}\, eV$

and $\Delta E_L \approx 6.59 \times 10^{-8}\, eV$ also.

$\Delta E_T = 1.32 \times 10^{-7}\, eV$ is the uncertainty in the Hα transition energy of $1.9 eV$.

5-37.

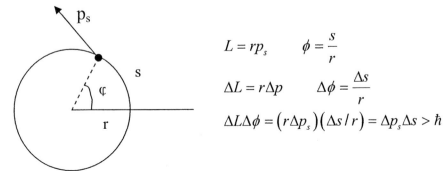

$L = rp_s \qquad \phi = \dfrac{s}{r}$

$\Delta L = r\Delta p \qquad \Delta \phi = \dfrac{\Delta s}{r}$

$\Delta L \Delta \phi = \left(r\Delta p_s\right)\left(\Delta s / r\right) = \Delta p_s \Delta s > \hbar$

(Problem 5-37 continued)

In the Bohr model, $L = n\hbar$ and may be known to within $\Delta L \approx 0.1\hbar$.

Then $\Delta\phi > \hbar/(0.1\hbar) = 10\,rad$. This exceeds one revolution, so that ϕ is completely

unknown.

5-41. (a) $E^2 = p^2c^2 + m^2c^4$ $\qquad E = hf = \hbar\omega \qquad p = h/\lambda = \hbar/k \qquad \hbar^2\omega^2 = \hbar^2k^2c^2 + m^2c^4$

$$v = \frac{\omega}{k} = \frac{\hbar\omega}{\hbar k} = \frac{\sqrt{\hbar^2k^2c^2 + m^2c^4}}{\hbar k} = c\sqrt{1 + m^2c^2/\hbar^2k^2} > c$$

(b) $v_s = \dfrac{d\omega}{dk} = \dfrac{d}{dk}\sqrt{\dfrac{k^2c^2 + m^2c^4}{\hbar k}} = \dfrac{c^2k}{\sqrt{\dfrac{k^2c^2 + m^2c^4}{\hbar^2}}}$

$$= \frac{c^2k}{\omega} = \frac{c^2\hbar k}{\hbar\omega} = \frac{c^2p}{E} = u \quad \text{(by Equation 2-41)}$$

5-45. (a) For proton: $E_1 = \dfrac{(hc)^2}{8m_pc^2L^2}$ from Problem 5-44.

$$E_1 = \frac{(1240\,MeV\bullet fm)^2}{8(938\,MeV)(1\,fm)^2} = 205\,MeV \text{ and } E_n = 205n^2\,MeV$$

$\therefore \quad E_2 = 820\,MeV$ and $E_3 = 1840\,MeV$

(b) For $n = 2 \to n = 1$ transition, $\lambda = \dfrac{hc}{\Delta E} = \dfrac{1240\,MeV\bullet fm}{615\,MeV} = 2.02\,fm$

(c) For $n = 3 \to n = 2$ transition, $\lambda = \dfrac{hc}{\Delta E} = \dfrac{1240\,MeV\bullet fm}{1020\,MeV} = 1.22\,fm$

(d) For $n = 3 \to n = 1$ transition, $\lambda = \dfrac{hc}{\Delta E} = \dfrac{1240\,MeV\bullet fm}{1635\,MeV} = 0.76\,fm$

5-49. (a) $\lambda = h/p$ The electrons are not moving at relativistic speeds, so

$$\lambda = h/mv = 6.63\times10^{-34}\,J\bullet s/(9.11\times10^{-31}\,kg)(3\times10^6\,m/s) = 2.43\times10^{-19}\,m = 0.243\,nm$$

(Problem 5-49 continued)

(b) The energy, momentum, and wavelength of the two photons are equal.

$$E = \frac{1}{2}mv^2 + mc^2 = \frac{1}{2}mc^2\left(v^2/c^2\right) + mc^2 = mc^2\left[\frac{1}{2}\left(v^2/c^2\right)+1\right]$$

$$= 0.511\times10^6\,eV\left[\frac{1}{2}\left(3\times10^6\right)/\left(3\times10^8\right)^2+1\right] \approx 0.511 MeV$$

(c) $p = E/c = 0.511 MeV/c$

(d) $\lambda = hc/E = 1240\,eV\bullet nm/0.511\times10^6\,eV = 2.43\times10^{-3}\,nm$

5-53. $\quad \frac{1}{2}m\overline{v^2} = \frac{3}{2}kT$

$$v_{rms} = \sqrt{\frac{3kT}{m}} = \left[\frac{3\left(1.381\times10^{-23}\,J/K\right)\left(300K\right)}{56u\left(1.66\times10^{-27}\,kg/u\right)}\right]^{1/2} = 366 m/s$$

$$f' = f_o\left(1+v/c\right) \quad \rightarrow \quad hf' = hf_o\left(1+v/c\right)$$

$$\Delta E = hf' - hf_o = hf_o v/c = \frac{\left(1eV\right)\left(366m/s\right)}{3.0\times10^8\,m/s} = 1.2\times10^{-6}\,eV$$

This is about 12 times the natural line width.

$$\Delta E = hf_o v/c = \frac{\left(10^6\,eV\right)\left(366m/s\right)}{3.0\times10^8\,m/s} = 1.2eV$$

This is over 10^7 times the natural line width.

Chapter 6 – The Schrödinger Equation

6-1. $\dfrac{d\Psi}{dx} = kAe^{kx-\omega t} = k\Psi$ and $\dfrac{d^2\Psi}{dx^2} = k^2\Psi$

Also, $\dfrac{d\Psi}{dt} = -\omega\Psi$. The Schrödinger equation is then, with these substitutions,

$-\hbar^2 k^2\Psi/2m + V\Psi = -i\hbar\omega\Psi$. Because the left side is real and the right side is a pure Imaginary number, the proposed Ψ does not satisfy Schrödinger's equation.

6-5. (a) $\Psi(x,t) = A\sin(kx - \omega t)$

$\dfrac{\partial\Psi}{\partial t} = -\omega A\cos(kx - \omega t)$

$i\hbar\dfrac{\partial\Psi}{\partial t} = -i\hbar\omega A\cos(kx - \omega t)$

$\dfrac{\partial^2\Psi}{\partial x^2} = -k^2 A\sin(kx - \omega t)$

$\dfrac{-\hbar^2}{2m}\dfrac{\partial^2\Psi}{\partial x^2} = \dfrac{-\hbar k^2 A}{2m}\sin(kx - \omega t) \neq i\hbar\dfrac{\partial\Psi}{\partial t}$

(b) $\Psi(x,t) = A\cos(kx - \omega t) + iA\sin(kx - \omega t)$

$i\hbar\dfrac{\partial\Psi}{\partial t} = i\hbar\omega A\sin(kx - \omega t) - i^2\hbar\omega A\cos(kx - \omega t)$

$\qquad = \hbar\omega A\cos(kx - \omega t) + i\hbar\omega A\sin(kx - \omega t)$

$-\dfrac{\hbar^2}{2m}\dfrac{\partial^2\Psi}{\partial x^2} = \dfrac{\hbar^2 k^2 A}{2m}\cos(kx - \omega t) + \dfrac{\hbar^2 i k^2 A}{2m}\sin(kx - \omega t)$

$\qquad = \dfrac{\hbar^2 k^2}{2m}\left[A\cos(kx - \omega t) + iA\sin(kx - \omega t)\right]$

$\qquad = i\hbar\dfrac{\partial\Psi}{\partial t}$ if $\dfrac{\hbar^2 k^2}{2m} = \hbar\omega$ it does. (Equation 6-5 with $V = 0$)

6-9. (a) The ground state of an infinite well is $E_1 = h^2/8mL^2 = (hc)^2/8mc^2L^2$

$$\text{For } m = m_p, \ L = 0.1nm: \quad E_1 = \frac{(1240MeV \cdot fm)^2}{8(938.3 \times 10^6 eV)(0.1nm)^2} = 0.021eV$$

(b) For $m = m_p$, $L = 1fm$: $\quad E_1 = \dfrac{(1240MeV \cdot fm)^2}{8(938.3 \times 10^6 eV)(1fm)^2} = 205MeV$

6-13. (a) $\Delta x = 0.0001L = (0.0001)(10^{-2}m) = 10^{-6}m$

$$\Delta p = 0.0001p = (0.0001)(10^{-9}kg)(10^{-3}m/s) = 10^{-16}kg \cdot m/s$$

(b) $\dfrac{\Delta x \Delta p}{\hbar} = \dfrac{(10^{-6}m)(10^{-16}kg \cdot m/s)}{1.055 \times 10^{-34}J \cdot s} = 9 \times 10^{11}$

6-17. The uncertainty principle requires that $\overline{E} \geq \dfrac{\hbar^2}{2mL^2}$ for any particle in any one-dimensional

box of width L (Equation 5-28). For a particle in an infinite one-dimensional square well:

$E_n = \dfrac{n^2h^2}{8mL^2}$

For $n = 0$, then E_0 must be 0 since $\dfrac{h^2}{8mL^2} > 0$. This violates Equation 5-28 and, hence, the

exclusion principle.

6-21. $E = -dE/dL$ comes from the impulse-momentum theorem $F\Delta t = 2mv$ where $\Delta t \approx L/v$.

So, $F \sim mv^2/L \sim E/L$. Because $E_1 = h^2/8mL^2$, $dE/dL = -h^2/4mL^3$ where the minus

sign means "on the wall". So $F = h^2/4mL^3 = \dfrac{(6.63 \times 10^{-34}J \cdot s)^2}{4(9.11 \times 10^{-31}kg)(10^{-10}m)^3} = 1.21 \times 10^{-7}N$

The weight of an electron is $mg = 9.11 \times 10^{-31}kg(9.8m/s^2) = 8.9 \times 10^{-30}N$ which is

minuscule by comparison.

6-25. Because $E_1 = 0.5eV$ and for a finite well also $E_n \approx n^2 E_1$, then $n = 4$ is at about $8eV$, i.e., near the top of the well. Referring to Figure 6-14, $ka \approx 2\pi$.

$$ka = \frac{\sqrt{2mE}}{\hbar} \times \frac{L}{2} = \left(7.24 \times 10^9\, m^{-1} \times L\right) = 2\pi$$

$$L = 2\pi / 7.24 \times 10^9 = 8.7 \times 10^{-10}\, m = 0.87nm$$

6-29. For $n = 3$, $\psi_3 = (2/L)^{1/2} \sin(3\pi x/L)$

(a) $\langle x \rangle = \int_0^L x (2/L) \sin^2 (3\pi x/L)\, dx$

Substituting $u = 3\pi x/L$, then $x = Lu/3\pi$ and $dx = (L/3\pi)\, du$. The limits become:

$x = 0 \to u = 0$ and $x = L \to u = 3\pi$

$$\langle x \rangle = (2/L)(L/3\pi)(1/3\pi) \int_0^{3\pi} u \sin^2 u\, du$$

$$= (2/L)(L/3\pi)^2 \left[\frac{u^2}{4} - \frac{u \sin 2u}{4} - \frac{\cos 2u}{8} \right]_0^{3\pi}$$

$$= (2/L)(1/3\pi)^2 (3\pi)^2 / 4 = L/2$$

(b) $\langle x^2 \rangle = \int_0^L x^2 (2/L) \sin^2 (3\pi x/L)\, dx$

Changing the variable exactly as in (a) and noting that:

$$\int_0^{3\pi} u^2 \sin^2 u\, du = \left[\frac{u^3}{6} - \left(\frac{u^2}{4} - \frac{1}{8} \right) \sin 2u - \frac{u \cos 2u}{4} \right]_0^{3\pi}$$

We obtain $\langle x^2 \rangle = \left(\frac{1}{3} - \frac{1}{8} \pi^2 \right) L^2 = 0.320 L^2$

6-33. $\psi_0(x) = A_0 e^{-m\omega x^2/2\hbar}$ where $A_0 = (m\omega/\hbar\pi)^{1/4}$

$\langle x \rangle = \int\limits_{-\infty}^{+\infty} A_0^2 x e^{-m\omega x^2/\hbar} dx$ Letting $u^2 = m\omega x^2/\hbar$ and $x = (\hbar/m\omega)^{1/2} u$

$2u\,du = (m\omega/\hbar)(2x\,dx)$. And thus, $(m\omega/\hbar)^{-1} u\,du = x\,dx$; limits are unchanged.

$\langle x \rangle = A_0^2 (\hbar/m\omega) \int\limits_{-\infty}^{+\infty} u e^{-u^2} du = 0$ (Note that the symmetry of $V(x)$ would also tell us that

$\langle x \rangle = 0$.)

$\langle x^2 \rangle = \int\limits_{-\infty}^{+\infty} A_0^2 x^2 e^{-m\omega x^2/\hbar} dx$

$\quad = A_0^2 (\hbar/m\omega)^{3/2} \int\limits_{-\infty}^{+\infty} u^2 e^{-u^2} du = 2A_0^2 (\hbar/m\omega)^{3/2} \int\limits_{-\infty}^{+\infty} u^2 e^{-u^2} du$

$\quad = 2A_0^2 (\hbar/m\omega)^{3/2} \sqrt{\pi}/4 = (m\omega/\hbar\pi)^{1/2} (\hbar/m\omega)^{3/2} \sqrt{\pi}/2 = \hbar/(2m\omega)$

6-37. $\psi_1(x) = C_1 x e^{-m\omega x^2/2\hbar}$ (Equation 6-58)

(a) $\int\limits_{-\infty}^{+\infty} |\psi_1(x)|^2 dx = 1 = \int\limits_{-\infty}^{+\infty} |C_1|^2 x^2 e^{-m\omega x^2/\hbar} dx = |C_1|^2 \times 2I_2$

$\quad = |C_1|^2 \times 2 \times \frac{1}{4}\sqrt{\frac{\pi}{\lambda^3}}$ with $\lambda = m\omega/\hbar$

$\quad = |C_1|^2 \times \frac{1}{2}\sqrt{\frac{\pi\hbar^3}{m^3\omega^3}}$

$\quad C_1 = \left(\frac{4m^3\omega^3}{\pi\hbar^3}\right)^{1/4}$

(b) $\langle x \rangle = \int\limits_{-\infty}^{+\infty} x|\psi_1|^2 dx = \int\limits_{-\infty}^{+\infty} x^3 \left(\frac{4m^3\omega^3}{\pi\hbar^3}\right)^{1/2} e^{-m\omega x^2/\hbar} dx = 0$

(c) $\langle x^2 \rangle = \int\limits_{-\infty}^{+\infty} x^2 |\psi_1|^2 dx = \int\limits_{-\infty}^{+\infty} x^2 \left(\frac{4m^3\omega^3}{\pi\hbar^3}\right)^{1/2} e^{-m\omega x^2/\hbar} x^2 dx$

(Problem 6-37 continued)

$$= \left(\frac{4m^3\omega^3}{\pi\hbar^3}\right)^{1/2} \times 2I_4 = \left(\frac{4m^3\omega^3}{\pi\hbar^3}\right)^{1/2} \times 2 \times \frac{3}{8}\sqrt{\frac{\pi}{\lambda^5}} \quad \text{where } \lambda = m\omega/\hbar$$

$$= \frac{3}{2}\sqrt{\frac{m^3\omega^3}{\pi\hbar^3}}\sqrt{\frac{\pi\hbar^5}{m^5\omega^5}} = \frac{3}{2}\frac{\hbar}{m\omega}$$

(d) $\langle V(x)\rangle = \left\langle\frac{1}{2}m\omega^2 x^2\right\rangle = \frac{1}{2}m\omega^2\langle x^2\rangle = \frac{1}{2}m\omega^2 \times \frac{3}{2}\frac{\hbar}{m\omega} = \frac{3}{4}\hbar\omega$

6-41. (a) $\omega = 2\pi f = 2\pi/T = 2\pi/1.42s = 4.42 rad/s$

$E_0 = \frac{1}{2}\hbar\omega = 1.055\times10^{-34}J\cdot s(4.42rad/s)/2 = 2.33\times10^{-34}J$

499.9 mm

500.0 mm

0.1mm

A

(b) $A = \sqrt{(500.0)^2 - (499.9)^2} = 10mm$

$E = (n+1/2)\hbar\omega = 1/2m\omega^2 A^2$

$n+1/2 = 1/2(0.010kg)(4.42rad/s)(10^{-2}m)^2\big/1.055\times10^{-34}J\cdot s$

$= 2.1\times10^{28}$ or $n = 2.1\times10^{28}$

(c) $f = \omega/2\pi = 0.70Hz$

6-45. (a)

$E = 4eV$

$9eV = V_0$

$0.6\ nm = a$

$\alpha = \sqrt{2m(V_0 - E)}\ \hbar$

$= \sqrt{2(0.511\times10^6 eV/c^2)(eV)}/\hbar$

$= \sqrt{5.11\times10^6 eV}\ \frac{eV}{c}/\hbar$

$= \frac{2260eV}{197.3eV\cdot nm} = 11.46 nm^{-1}$

41

(Problem 6-45 continued)

and $\alpha a = 0.6nm \times 11.46nm^{-1} = 6.87$

Since αa is not $\ll 1$, use Equation 6-75:

The transmitted fraction

$$T = \left[1 + \frac{\sinh^2 \alpha a}{4(E/V_0)(1 - E/V_0)}\right]^{-1} = \left[1 + \left(\frac{81}{80}\right)\sinh^2(6.87)\right]^{-1}$$

Recall that $\sinh x = (e^x - e^{-x})/2$,

$$T = \left[1 + \frac{81}{80}\left(\frac{e^{6.87} - e^{-6.87}}{2}\right)^2\right]^{-1} = 4.3 \times 10^{-6} \text{ is the transmitted fraction.}$$

(b) Noting that the size of T is controlled by αa through the $\sinh^2 \alpha a$ and increasing T implies increasing E. Trying a few values, selecting $E = 4.5eV$ yields $T = 8.7 \times 10^{-6}$ or approximately twice the value in part (a).

6-49. $R = \dfrac{(k_1 - k_2)^2}{(k_1 + k_2)^2}$ and $T = 1 - R$ (Equations 6-68 and 6-70)

(a) For protons:

$$k_1 = \sqrt{2mc^2 E}/\hbar c = \sqrt{2(938MeV)(40MeV)}/197.3MeV \bullet fm = 1.388$$

$$k_2 = \sqrt{2mc^2(E - V_0)}/\hbar c = \sqrt{2(938MeV)(10MeV)}/197.3MeV \bullet fm = 0.694$$

$$R = \left(\frac{1.388 - 0.694}{1.388 + 0.694}\right)^2 = \left(\frac{0.694}{2.082}\right)^2 = 0.111 \quad \text{And } T = 1 - R = 0.889$$

(b) For electrons:

$$k_1 = 1.388\left(\frac{0.511}{938}\right)^{1/2} = 0.0324 \qquad k_2 = 0.694\left(\frac{0.511}{938}\right)^{1/2} = 0.0162$$

$$R = \left(\frac{0.0324 - 0.0162}{0.0324 + 0.0162}\right)^2 = 0.111 \quad \text{And } T = 1 - R = 0.889$$

No, the mass of the particle is not a factor. (We might have noticed that $\sqrt{m}$ could be canceled from each term.)

42

6-53. The $n = 2$ wave function is $\psi_2(x) = \sqrt{\dfrac{2}{L}}\sin\dfrac{2\pi x}{L}$ and the kinetic energy operator

$$\left(E_k\right)_{op} = -\frac{\hbar^2}{2m}\frac{\partial^2}{\partial x^2} \quad \text{Therefore,}$$

$$\langle E_k \rangle = \int_{-\infty}^{+\infty} \psi_2^*(x)\left(-\frac{\hbar^2}{2m}\frac{\partial^2}{\partial x^2}\right)\psi_2(x)\,dx$$

$$= \int_{-\infty}^{+\infty} \sqrt{\frac{2}{L}}\sin\frac{2\pi x}{L}\left(-\frac{\hbar^2}{2m}\frac{\partial^2}{\partial x^2}\right)\sqrt{\frac{2}{L}}\sin\frac{2\pi x}{L}\,dx$$

$$= \frac{2}{L}\left(-\frac{\hbar^2}{2m}\right)\int_0^L \sin\frac{2\pi x}{L}\left(\frac{2\pi}{L}\right)^2\left(-\sin\frac{2\pi x}{L}\right)dx$$

$$= \frac{2}{L}\left(\frac{\hbar^2}{2m}\right)\left(\frac{2\pi}{L}\right)^2\int_0^L \sin^2\frac{2\pi x}{L}\,dx$$

Let $\dfrac{2\pi x}{L} = y$, then $x = 0 \rightarrow y = 0$ and $x = L \rightarrow y = 2\pi$ and $\dfrac{2\pi dx}{L} = dy \rightarrow dx = \dfrac{L}{2\pi}dy$

Substituting above gives: $\langle E_k \rangle = \dfrac{2}{L}\left(\dfrac{\hbar^2}{2m}\right)\left(\dfrac{2\pi}{L}\right)^2\left(\dfrac{L}{2\pi}\right)\displaystyle\int_0^{2\pi}\sin^2 y\,dy$

$$\int_0^{2\pi}\sin^2 y\,dy = \left(\frac{y}{2}-\frac{\sin 2y}{4}\right)\Bigg|_0^{2\pi} = \left[\left(\frac{2\pi}{2}-0\right)-(0-0)\right] = \pi$$

Therefore, $\langle E_k \rangle = \dfrac{2}{L}\left(\dfrac{\hbar^2}{2m}\right)\left(\dfrac{2\pi}{L}\right)^2\left(\dfrac{L}{2\pi}\right)\pi = \dfrac{4\pi^2\hbar^2}{2mL^2} = \dfrac{h^2}{2mL^2}$

6-57. $T \approx 16\dfrac{E}{V_0}\left(1-\dfrac{E}{V_0}\right)e^{-2\alpha a}$ where $E = 10eV$, $V_0 = 25eV$, and $\alpha = 1nm$.

(a) $\alpha = \sqrt{2m(V_0-E)}\big/\hbar = \sqrt{2(m_0c^2)(V_0-E)}\big/\hbar c$

$$= \sqrt{2(0.511\times10^6\,eV)(15eV)}\big/197.3eV\cdot nm = 19.84nm^{-1}$$

And $\alpha a = (19.84nm^{-1})(1nm) = 19.84; \quad 2\alpha a = 29.68$

$$T \approx 16\left(\frac{10}{25}\right)\left(1-\frac{10}{25}\right)e^{-29.68} \approx 4.95\times10^{-13}$$

(Problem 6-57 continued)

(b) For $a = 0.1nm$: $\alpha a = (19.84nm^{-1})(0.1nm) = 1.984$

$$T \approx 16\left(\frac{10}{25}\right)\left(1 - \frac{10}{25}\right)e^{-2.968} \approx 0.197$$

6-61. (a) $\Delta p \Delta x \approx \hbar \rightarrow m\Delta v \Delta x \approx \hbar$

$$\Delta v \approx \hbar / m\Delta a = (1.055\times10^{-34}J\bullet s)/(9.11\times10^{-31})(10^{-12}m)$$

$$\Delta v \approx 1.6\times10^8 m/s = 0.39c$$

(b) The width of the well L is still an integer number of half wavelengths, $L = n(\lambda/2)$, and deBroglie's relation still gives: $L = nh/2p$. However, p is *not* given by:

$p = \sqrt{2mE_k}$, but by the relativistic expression: $p = \left[E^2 - (mc^2)^2\right]^{1/2}\Big/c.$

Substituting this yields: $L = \dfrac{nhc}{2\left[E^2 - (mc^2)^2\right]^{1/2}} \rightarrow E^2 - (mc^2)^2 = (nhc/2L)^2$

$$E_n = \left[\left(\frac{nhc}{2L}\right)^2 + (mc^2)^2\right]^{1/2}$$

(c) $E_1 = \left[\left(\dfrac{hc}{4L^2}\right)^2 + (mc^2)^2\right]^{1/2} = \left[\dfrac{(1240eV\bullet nm)^2}{4(10^{-3}nm)^2} + (0.511\times10^6 eV)^2\right]^{1/2} = 8.03\times10^5 eV$

(d) Nonrelativistic:

$$E_1 = \frac{h^2}{8mL^2} = \frac{(hc)^2}{8(mc^2)L^2} = \frac{(1240eV\bullet nm)^2}{8(0.511\times10^6 eV)(10^{-3}nm)^2} = 3.76\times10^5 eV$$

E_1 computed in (c) is 2.14 times the nonrelativistic value.

Chapter 7 – Atomic Physics

7-1. $E_{n_1 n_2 n_3} = \dfrac{\hbar^2 \pi^2}{2mL^2}\left(n_1^2 + n_2^2 + n_3^2\right)$ (Equation 7-4)

$E_{311} = \dfrac{\hbar^2 \pi^2}{2mL^2}\left(3^2 + 1^2 + 1^2\right) = 11E_0$ where $E_0 = \dfrac{\hbar^2 \pi^2}{2mL^2}$

$E_{222} = E_0\left(2^2 + 2^2 + 2^2\right) = 12E_0$ and $E_{321} = E_0\left(3^2 + 2^2 + 1^2\right) = 14E_0$

The 1st, 2nd, 3rd, and 5th excited states are degenerate.

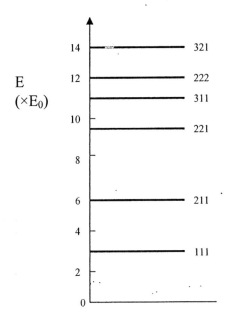

7-5. $E_{n_1 n_2 n_3} = \dfrac{\hbar^2 \pi^2}{2m}\left(\dfrac{n_1^2}{L_1^2} + \dfrac{n_2^2}{\left(2L_1\right)^2} + \dfrac{n_3^2}{\left(4L_1\right)^2}\right) = \dfrac{\hbar^2 \pi^2}{2mL_1^2}\left(n_1^2 + \dfrac{n_2^2}{4} + \dfrac{n_3^2}{16}\right)$ (from Equation 7-5)

$E_0 = \left(n_1^2 + \dfrac{n_2^2}{4} + \dfrac{n_3^2}{16}\right)$ where $E_0 = \dfrac{\hbar^2 \pi^2}{2mL_1^2}$

(Problem 7-5 continued)

(a)

n_1	n_2	n_3	$E\left(\times E_0\right)$
1	1	1	1.313
1	1	2	1.500
1	1	3	1.813
1	2	1	2.063
1	1	4	2.250
1	2	2	2.250
1	2	3	2.563
1	1	5	2.813
1	2	4	3.000
1	1	6	3.500

(b) 1,1,4 and 1,2,2

7-9. (a) For $n = 3$, $\ell = 0, 1, 2$

  (b) For $\ell = 0$, $m = 0$. For $\ell = 1$, $m = -1, 0, +1$. For $\ell = 2$, $m = -2, -1, 0, +1, +2$.

  (c) There are nine different m-states, each with two spin states, for a total of 18 states for $n = 3$.

7-13. $L^2 = L_x^2 + L_y^2 + L_z^2 \rightarrow L_x^2 + L_y^2 = L^2 - L_z^2 = \ell(\ell+1)\hbar^2 - (m\hbar)^2 = \left(6 - m^2\right)\hbar^2$

  (a) $\left(L_x^2 + L_y^2\right)_{min} = \left(6 - 2^2\right)\hbar^2 = 2\hbar^2$

  (b) $\left(L_x^2 + L_y^2\right)_{max} = \left(6 - 0^2\right)\hbar^2 = 6\hbar^2$

  (c) $L_x^2 + L_y^2 = (6-1)\hbar^2 = 5\hbar^2$ L_x and L_y cannot be determined separately.

  (d) $n = 3$

7-17. (a) $6f$ state: $n = 6$, $\ell = 3$

(b) $E_6 = -13.6eV / n^2 = -13.6eV / 6^2 = -0.38eV$

(c) $L = \sqrt{\ell(\ell+1)}\,\hbar = \sqrt{3(3+1)}\,\hbar = \sqrt{12}\,\hbar = 3.65 \times 10^{-34}\,J\bullet s$

(d) $L_z = m\hbar$ $L_z = -3\hbar, -2\hbar, -1\hbar, 0, 1\hbar, 2\hbar, 3\hbar$

7-21. $P(r) = Cr^2 e^{-2Zr/a_0}$ For $P(r)$ to be a maximum,

$$\frac{dP}{dt} = C\left[r^2 \left(-\frac{2Z}{a_0} \right) e^{-2Zr/a_0} + 2re^{-2Zr/a_0} \right] = 0 \;\rightarrow\; C \times \frac{2Zr}{a_0}\left(\frac{a_0}{Z} - r \right) e^{-2Zr/a_0} = 0$$

This condition is satisfied with $r = 0$ or $r = a_0/Z$. For $r = 0$, $P(r) = 0$ so the maximum

$P(r)$ occurs for $r = a_0/Z$.

7-25. $\psi_{200} = \frac{1}{\sqrt{32\pi}}\left(\frac{1}{a_0} \right)^{3/2} \left(2 - \frac{r}{2a_0} \right) e^{-r/2a_0}$ ($Z = 1$ for hyrdogen)

(a) At $r = a_0$, $\psi_{200} = \frac{1}{\sqrt{32\pi}}\left(\frac{1}{a_0^3} \right)(2-1)e^{-1/2} = \frac{0.606}{\sqrt{32\pi}}\left(\frac{1}{a_0} \right)^{3/2}$

(b) At $r = a_0$, $|\psi_{200}|^2 = \frac{1}{\sqrt{32\pi}}\left(\frac{1}{a_0^3} \right)e^{-1} = \frac{0.368}{32\pi}\frac{1}{a_0^3}$

(c) At $r = a_0$, $P(r) = |\psi_{200}|^2 (4\pi r^2) = \frac{4}{32\pi}\frac{0.368a_0^2}{a_0^3} = \frac{0.368}{8a_0}$

7-29. (a) Every increment of charge follows a circular path of radius R and encloses an area

πR^2, so the magnetic moment is the total current times this area. The entire charge

Q rotates with frequency $f = \omega/2\pi$, so the current is

$i = Qf = q\omega/2\pi$

$\mu = iA = (Q\omega/2\pi)(\pi R^2) = Q\omega R^2/2$

(Problem 7-29 continued)

$$L = I\omega = \frac{1}{2}MR^2\omega$$

$$g = \frac{2M\mu}{QL} = \frac{2MQ\omega R^2 / 2}{QMR^2\omega / 2} = 2$$

    (b) The entire charge is on the equatorial ring, which rotates with frequency $f = \omega / 2\pi$.

$$i = Qf = Q\omega / 2\pi$$

$$\mu = iA = \left(Q\omega / 2\pi\right)\left(\pi R^2\right) = Q\omega R^2 / 2$$

$$g = \frac{2M\mu}{QL} = \frac{2MQ\omega R^2 / 2}{QMR^2\omega / 5} = 5/2 = 2.5$$

7-33.   (a) There should be four lines corresponding to the four m_J values $-3/2$, $-1/2$, $+1/2$, $+3/2$.

     (b) There should be three lines corresponding to the three m_ℓ values -1, 0, $+1$.

7-37.   (a) $d_{5/2}$

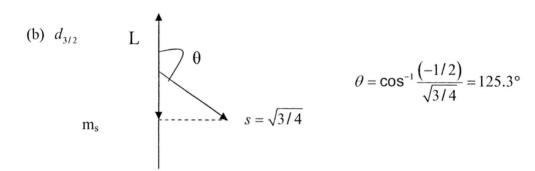

$$\theta = \cos^{-1}\frac{1/2}{\sqrt{3/4}} = 54.7°$$

     (b) $d_{3/2}$

$$\theta = \cos^{-1}\frac{(-1/2)}{\sqrt{3/4}} = 125.3°$$

7-41. $\psi_{12} = \psi(x_1, x_2) = C \sin \dfrac{\pi x_1}{L} \sin \dfrac{2\pi x_2}{L}$ Substituting into Equation 7-57 with $V = 0$,

$$-\frac{\hbar^2}{2m}\left(\frac{\partial^2 \psi_{12}}{\partial x_1^2} + \frac{\partial^2 \psi_{12}}{\partial x_2^2}\right) = \left(\frac{\hbar^2}{2m}\right)(1+4)\left(\frac{\pi^2}{L^2}\right)\psi_{12} = E\psi_{12}$$

Obviously, ψ_{12} is a solution if $E = \dfrac{5\hbar^2 \pi^2}{2mL^2}$

7-45.   Using Figure 7-34:

Sn ($Z = 50$)

   $1s^2\, 2s^2\, 2p^6\, 3s^2\, 3p^6\, 3d^{10}\, 4s^2\, 4p^6\, 4d^{10}\, 5s^2\, 5p^2$

Nd ($Z = 60$)

   $1s^2\, 2s^2\, 2p^6\, 3s^2\, 3p^6\, 3d^{10}\, 4s^2\, 4p^6\, 4d^{10}\, 5s^2\, 5p^6\, 4f^4\, 6s^2$

Yb ($Z = 70$)

   $1s^2\, 2s^2\, 2p^6\, 3s^2\, 3p^6\, 3d^{10}\, 4s^2\, 4p^6\, 4d^{10}\, 4f^{14}\, 5s^2\, 5p^6\, 6s^2$

Comparison with Appendix C.

Sn: agrees

Nd: $5p^6$ and $4f^4$ are in reverse order

Yb: agrees

7-49.   (a) Fourteen electrons, so $Z = 14$. Element is silicon.

   (b) Twenty electrons, so $Z = 20$. Element is calcium.

7-53.   Similar to H: Li, Rb, Ag, and Fr.   Similar to He: Ca, Ti, Cd, Ba, Hg, and Ra.

7-57. $\Delta j = \pm 1, 0$ $\left(\text{no } j = 0 \rightarrow j = 0\right)$ (Equation 7-66)

The four states are $^2P_{3/2}$, $^2P_{1/2}$, $^2D_{5/2}$, $^2D_{3/2}$

Transition	$\Delta\ell$	Δj	Comment
$D_{5/2} \rightarrow P_{3/2}$	-1	-1	allowed
$D_{5/2} \rightarrow P_{1/2}$	-1	-2	j - forbidden
$D_{3/2} \rightarrow P_{3/2}$	-1	0	allowed
$D_{3/2} \rightarrow P_{1/2}$	-1	-1	allowed

7-61. (a) $\Delta E = \dfrac{e\hbar}{2m}B = \left(5.79 \times 10^{-5}\,eV/T\right)\left(0.05T\right) = 2.90 \times 10^{-6}\,eV$

(b) $\left|\Delta\lambda\right| = \dfrac{\lambda^2}{hc}\Delta E = \dfrac{\left(579.07nm\right)^2\left(2.90 \times 10^{-6}\,eV\right)}{1240eV\cdot nm} = 7.83 \times 10^{-4}\,nm$

(c) The smallest measurable wavelength change is larger than this by the ratio $0.01nm/7.83 \times 10^{-4}\,nm = 12.8$. The magnetic field would need to be increased by this same factor because $B \propto \Delta E \propto \Delta\lambda$. The necessary field would be 0.638T.

7-65. (a) $\left|F_z\right| = m_s g_L \mu_B \left(dB/dz\right)$ (From Equation 7-51)

From Newton's 2nd law, $\left|F_z\right| = m_H a_z = m_s g_L \mu_B \left(dB/dz\right)$

$a_z = m_s g_L \left(dB/dz\right)/m_H = \left(1/2\right)\left(1\right)\left(9.27 \times 10^{-24}\,J/T\right)\left(600T/m\right)/\left(1.67 \times 10^{-27}\,kg\right)$

$= 1.67 \times 10^6\,m/s^2$

(b) At $14.5km/s = v = 1.45 \times 10^4\,m/s$, the atom takes $t_1 = 0.75m/\left(1.45 \times 10^4\,m/s\right)$

$= 5.2 \times 10^{-5}\,s$ to traverse the magnet. In that time, its z deflection will be:

$z_1 = \left(1/2\right)\left(a_z\right)t_1^2 = \left(1/2\right)\left(1.67 \times 10^6\,m/s^2\right)\left(5.2 \times 10^{-5}\,s\right)^2 = 2.26 \times 10^{-3}\,m = 2.26mm$

Its v_z velocity component as it leaves the magnet is $v_z = a_z t_1$ and its additional z deflection before reaching the detector 1.25m away will be:

(Problem 7-65 continued)

$$z_2 = v_z t_2 = (a_z t_1)\left(1.25m\left/\left[1.45\times10^4 m/s\right]\right.\right)$$

$$= \left(1.67\times10^6 m/s^2\right)\left(5.2\times10^{-5}s\right)(1.25)\left/\left(1.45\times10^4 m/s\right)\right.$$

$$= 7.49\times10^{-3}m = 7.49mm$$

Each line will be deflected $z_1 + z_2 = 9.75mm$ from the central position and, thus, separated by a total of $19.5mm = 1.95cm$.

7-69.   (a) $g = 1 + \dfrac{j(j+1)+s(s+1)-\ell(\ell+1)}{2j(j+1)}$ (Equation 7-73)

For $^2P_{1/2}$: $j = 1/2$, $\ell = 1$, and $s = 1/2$

$$g = 1 + \frac{1/2(1/2+1)+1/2(1/2+1)-1(1+1)}{2\times1/2(1/2+1)} = 1 + \frac{3/4+3/4-2}{3/2} = 2/3$$

For $^2S_{1/2}$: $j = 1/2$, $\ell = 0$, and $s = 1/2$

$$g = 1 + \frac{1/2(1/2+1)+1/2(1/2+1)-0}{2\times1/2(1/2+1)} = 1 + \frac{3/4+3/4}{3/2} = 2$$

The $^2P_{1/2}$ levels shift by:

$$\Delta E = g m_j \mu_B B = \frac{2}{3}\left(\pm\frac{1}{2}\right)\mu_B B = \pm\frac{1}{3}\mu_B B \qquad \text{(Equation 7-72)}$$

The $^2S_{1/2}$ levels shift by:

$$\Delta E = g m_j \mu_B B = 2\left(\pm\frac{1}{2}\right)\mu_B B = \pm\mu_B B$$

To find the transition energies, tabulate the several possible transitions and the corresponding energy values (let E_p and E_s be the $B = 0$ unsplit energies of the two states.):

(Problem 7-69 continued)

<u>Transition</u>   <u>Energy</u>

$P_{1/2,1/2} \rightarrow S_{1/2,1/2}$ $\left(E_p + \dfrac{1}{3}\mu_B B\right) - \left(E_s + \mu_B B\right) = \left(E_p - E_s\right) - \dfrac{2}{3}\mu_B B$

$P_{1/2,-1/2} \rightarrow S_{1/2,1/2}$ $\left(E_p - \dfrac{1}{3}\mu_B B\right) - \left(E_s + \mu_B B\right) = \left(E_p - E_s\right) - \dfrac{4}{3}\mu_B B$

$P_{1/2,1/2} \rightarrow S_{1/2,-1/2}$ $\left(E_p + \dfrac{1}{3}\mu_B B\right) - \left(E_s - \mu_B B\right) = \left(E_p - E_s\right) + \dfrac{4}{3}\mu_B B$

$P_{1/2,-1/2} \rightarrow S_{1/2,-1/2}$ $\left(E_p - \dfrac{1}{3}\mu_B B\right) - \left(E_s - \mu_B B\right) = \left(E_p - E_s\right) + \dfrac{2}{3}\mu_B B$

Thus, there are four different photon energies emitted. The energy or frequency spectrum would appear as below (normal Zeeman spectrum shown for comparison).

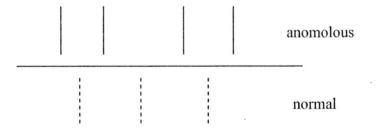

(b) For $^2P_{3/2}$: $j = 3/2$, $\ell = 1$, and $s = 1/2$

$$g = 1 + \frac{3/2(3/2+1) + 1/2(1/2+1) - 1(1+1)}{2 \times 3/2(3/2+1)} = 1 + \frac{15/4 + 3/4 - 2}{30/4} = 4/3$$

These levels shift by:

$$\Delta E = g m_j \mu_B B = \frac{4}{3}\left(\pm\frac{1}{2}\right)\mu_B B = \pm\frac{2}{3}\mu_B B \qquad \Delta E = \frac{4}{3}\left(\pm\frac{3}{2}\right)\mu_B B = \pm 2\mu_B B$$

Tabulating the transitions as before:

<u>Transition</u>   <u>Energy</u>

$P_{3/2,3/2} \rightarrow S_{1/2,1/2}$ $\left(E_p + 2\mu_B B\right) - \left(E_s + \mu_B B\right) = \left(E_p - E_s\right) + \mu_B B$

$P_{3/2,3/2} \rightarrow S_{1/2,-1/2}$ forbidden, $\Delta m_j = 2$

$P_{3/2,1/2} \rightarrow S_{1/2,1/2}$ $\left(E_p - \dfrac{2}{3}\mu_B B\right) - \left(E_s + \mu_B B\right) = \left(E_p - E_s\right) - \dfrac{1}{3}\mu_B B$

(Problem 7-69 continued)

Transition	Energy
$P_{3/2,1/2} \rightarrow S_{1/2,-1/2}$	$\left(E_p + \dfrac{2}{3}\mu_B B\right) - \left(E_s - \mu_B B\right) = \left(E_p - E_s\right) + \dfrac{5}{3}\mu_B B$
$P_{3/2,-1/2} \rightarrow S_{1/2,1/2}$	$\left(E_p - \dfrac{2}{3}\mu_B B\right) - \left(E_s + \mu_B B\right) = \left(E_p - E_s\right) - \dfrac{5}{3}\mu_B B$
$P_{3/2,-1/2} \rightarrow S_{1/2,-1/2}$	$\left(E_p - \dfrac{2}{3}\mu_B B\right) - \left(E_s - \mu_B B\right) = \left(E_p - E_s\right) + \dfrac{1}{3}\mu_B B$
$P_{3/2,-3/2} \rightarrow S_{1/2,1/2}$	forbidden, $\Delta m_j = 2$
$P_{3/2,-3/2} \rightarrow S_{1/2,-1/2}$	$\left(E_p - 2\mu_B B\right) - \left(E_s - \mu_B B\right) = \left(E_p - E_s\right) - \mu_B B$

There are six different photon energies emitted (two transitions are forbidden); their spectrum looks as below:

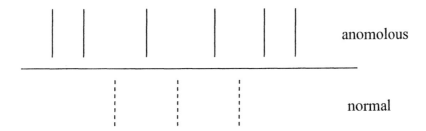

7-73. (a) $J = L + S$ $\mu = -\mu_B (L + 2S)/\hbar$ (Equation 7-71)

$$\mu_J = \frac{\mu \cdot J}{J} = \frac{[-\mu_B (L+2S)/\hbar] \times [L + S]}{J} \equiv -\frac{\mu_B}{\hbar J}(L \cdot L + 2S \cdot S + 3S \cdot L)$$

$$= -\frac{\mu_B}{\hbar J}\left(L^2 + 2S^2 + 3S \cdot L\right)$$

(b) $J^2 = J \cdot J = (L+S) \cdot (L+S) = L \cdot L + S \cdot S + 2S \cdot L$ $\therefore S \cdot L = \dfrac{1}{2}\left(J^2 - L^2 - S^2\right)$

(c) $\mu_J = -\dfrac{\mu_B}{\hbar J}\left[L^2 + 2S^2 + \dfrac{3}{2}\left(J^2 - L^2 - S^2\right)\right] = -\dfrac{\mu_B}{2\hbar J}\left(3J^2 + S^2 - L^2\right)$

(d) $\mu_z = \mu_J \dfrac{J_z}{J} = -\dfrac{\mu_B}{2\hbar J}\left(3J^2 + S^2 - L^2\right)\dfrac{J_z}{J} = -\dfrac{\mu_B}{2\hbar J}\left(3J^2 + S^2 - L^2\right)$

(Problem 7-73 continued)

$$= -\mu_B \left(1 + \frac{J^2 + S^2 - L^2}{2J^2} \right) \frac{J_z}{\hbar}$$

(e) $\Delta E = -\mu_z B$ (Equation 7-69)

$$= +\mu_B B \left[1 + \frac{j(j+1) + s(s+1) - \ell(\ell+1)}{2j(j+1)} \right] m_j$$

$$= g m_j \mu_B B \quad \text{(Equation 7-72)}$$

where $g = \left[1 + \dfrac{j(j+1) + s(s+1) - \ell(\ell+1)}{2j(j+1)} \right]$ (Equation 7-73)

Chapter 8 – Statistical Physics

8-1. (a) $v_{rms} = \sqrt{\dfrac{3RT}{M}} = \left[\dfrac{3(8.31J/mole\bullet K)(300K)}{2(.0079\times10^{-3}kg/mole)}\right]^{1/2} = 1930m/s$

(b) $T = \dfrac{Mv_{rms}^2}{3R} = \dfrac{2(1.0079\times10^{-3}kg/mole)(11.2\times10^3 m/s)^2}{3(8.31J/mole\bullet K)} = 1.01\times10^4 K$

8-5. (a) $E_K = n\times\dfrac{3}{2}RT = (1\ mole)\dfrac{3}{2}(8.31J/mole\bullet K)(273) = 3400J$

(b) One mole of any gas has the same translational energy at the same temperature.

8-9. $n(v)dv = 4\pi N\left(\dfrac{m}{2\pi kT}\right)^{3/2} v^2 e^{-mv^2/2kT}dv$ (Equation 8-8)

$\dfrac{dn}{dv} = A\left[v^2\left(-\dfrac{2vm}{2kT}\right)+2v\right]e^{-mv^2/2kT}$ The v for which $dn/dv=0$ is v_m.

$A\left[-\dfrac{2mv^3}{2kT}+2v\right]e^{-mv^2/2kT} = 0$

Because A = constant and the exponential term is only zero for $v\to\infty$, only the quantity

in [] can be zero, so $-\dfrac{2mv^3}{2kT}+2v = 0$

or $v^2 = \dfrac{2kT}{m} \to v_m = \sqrt{\dfrac{2kT}{m}}$ (Equation 8-9)

8-13. There are two degrees of freedom, therefore,

$C_v = 2(R/2) = R,\quad C_p = R+R = 2R,$ and $\gamma = 2R/R = 2.$

8-17. For hydrogen: $E_n = -\dfrac{mk^2e^4}{2\hbar^2}\dfrac{1}{n^2} = -\dfrac{13.605687}{n^2}eV$ using values of the constants accurate to six decimal places.

$E_1 = -13.605687eV$

$E_2 = -3.401422eV \qquad E_2 - E_1 = 10.204265eV$

$E_3 = -1.511743eV \qquad E_3 - E_1 = 12.093944eV$

(a) $\dfrac{n_2}{n_1} = \dfrac{g_2}{g_1}e^{-(E_2-E_1)/kT} = \dfrac{8}{2}e^{-10.20427/0.02586} = 4e^{-395} = 4\times10^{-172} \approx 0$

$\dfrac{n_3}{n_1} = \dfrac{g_3}{g_1}e^{-(E_3-E_1)/kT} = \dfrac{18}{2}e^{-12.09394/0.02586} = 9e^{-468} = 9\times10^{-203} \approx 0$

(b) $\dfrac{n_2}{n_1} = 0.01 = 4e^{-10.20427/kT} \quad \rightarrow \quad e^{-10.20427/kT} = 0.0025$

$-10.20427/kT = \ln(0.0025) = -5.99146$

$T = \dfrac{10.20427eV}{(5.99146)(8.61734\times10^{-5}eV\cdot K)} = 19,760K$

(c) $\dfrac{n_3}{n_1} = 9e^{-12.09394/(8.61734\times10^{-5})(19,760)} = 0.00742 = 0.7\%$

8-21. Assuming the gases are ideal gases, the pressure is given by: $P = \dfrac{2}{3}\dfrac{N\langle E\rangle}{V}$ for classical, FD, and BE particles. P_{FD} will be highest due to the exclusion principle, which, in effect, limits the volume available to each particle so that each strikes the walls more frequently than the classical particles. On the other hand, P_{BE} will be lowest, because the particles tend to be in the same state, which in effect, is like classical particles with a mutual attraction, so they strike the walls less frequently.

8-25. For small values of α, $e^\alpha = 1 + \alpha + (\alpha^2/2!) + \cdots$ and $N_0 = \dfrac{1}{e^\alpha - 1} \rightarrow N_0(e^\alpha - 1) = 1$

which for small α values becomes: $N_0(1 + \alpha + \cdots - 1) = N_0\alpha = 1$ or $N_0 = \dfrac{1}{\alpha}$

8-29. $C_V = 3N_A k \left(\dfrac{hf}{kT}\right)^2 \dfrac{e^{hf/kT}}{\left(e^{hf/kT}-1\right)^2}$ As $T \to \infty$, hf/kT gets small and

$e^{hf/kT} \approx 1 + hf/kT + \cdots$

$C_V = 3N_A k \left(\dfrac{hf}{kT}\right)^2 \dfrac{\left(1+hf/kT+\cdots\right)}{\left(hf/kT\right)^2} \approx 3N_A k = 3N_A \left(R/N_A\right) = 3R$

The rule of Dulong and Petit.

8-33. The photon gas has the most states available, since any number of photons may be in the ground state. In contrast, at $T = 1K$ the electron gas' available states are limited to those within about $2kT = (2)(8.62 \times 10^{-5} eV \cdot K)(1K) = 1.72 \times 10^{-4} eV$ of the Fermi level. All other states are either filled, hence unavailable, or higher than kT above the Fermi level, hence not accessible.

8-37. (a) $f_{FD}(E) = \dfrac{1}{e^{(E-E_F/kT)}+1}$ (Equation 8-68)

$= \dfrac{1}{e^{(E-E_F)/0.1E_F}+1} = \dfrac{1}{e^{10(E-E_F)/E_F}+1}$

(b) $f_{FD}(E) = \dfrac{1}{e^{(E-E_F)/0.5E_F}+1} = \dfrac{1}{e^{2(E-E_F)/E_F}+1}$

Al

8-41. $\left\langle E_{K(escape)}\right\rangle = \frac{1}{2}m\left\langle v^2_{escape}\right\rangle = \int_0^\infty \left(\frac{1}{2}mv^2\right)F(v)\,dv$

$$= \frac{\int_0^\infty \left(\frac{1}{2}mv^2\right)v^3 e^{-mv^2/2kT}\,dv}{\int_0^\infty v^3 e^{-mv^2/2kT}\,dv} = \frac{1}{2}m\frac{I_5}{I_3}$$

$$= \frac{1}{2}m\frac{\lambda^{-3}}{\lambda^{-2}/2} = \frac{m}{\lambda} = m\frac{2kT}{m} = 2kT \quad \text{where } \lambda = \frac{m}{2kT}$$

8-45. $N = e^{-\alpha}\frac{4\pi(2m_e)^{3/2}V}{h^3}\int_0^\infty E^{1/2}e^{-E/kT}\,dE$ (Equation 8-43)

Considering the integral, we change the variable: $E/kT = u^2$, then

$E = kTu^2$, $E^{1/2} = (kT)^{1/2}u$, and $dE = kT(2u)\,du$. So,

$$\int_0^\infty E^{1/2}e^{-E/kT}\,dE = 2(kT)^{3/2}\int_0^\infty u^2 e^{-u^2}\,du$$

The value of the integral (from tables) is $\sqrt{\pi}/4$, so

$$N = e^{-\alpha}\frac{4\pi(2m_e)^{3/2}V}{h^3}\frac{2(kT)^{3/2}\sqrt{\pi}}{4} \quad \text{or} \quad e^{\alpha} = \frac{2(2m_e\pi kT)^{3/2}V}{Nh^3}, \text{ which is Equation 8-44.}$$

Chapter 9 – Molecular Structure and Spectra

9-1. (a) $1\dfrac{eV}{molecule}=\left(1\dfrac{eV}{molecule}\right)\left(\dfrac{1.609\times10^{-19}J}{eV}\right)\left(\dfrac{6.022\times10^{23}\,molecules}{mole}\right)$

$$=\left(96472\dfrac{J}{mole}\right)\left(\dfrac{1cal}{4.184J}\right)=23057\dfrac{cal}{mole}=23.06\dfrac{kcal}{mole}$$

(b) $E_d=\left(4.27\dfrac{eV}{molecule}\right)\left(\dfrac{23.06kcal\,/\,mole}{1eV\,/\,molecule}\right)=98.5kcal\,/\,mole$

(c) $E_d=\left(106\dfrac{eV}{molecule}\right)\left(\dfrac{1eV\,/\,molecule}{96.47kJ\,/\,mole}\right)=1.08eV\,/\,molecule$

9-5. (a) Total potential energy: $U(r)=-\dfrac{ke^2}{r}+E_{ex}+E_{ion}$ (Equation 9-1)

attractive part of $U(r_0)=-\dfrac{ke^2}{r_0}=-\dfrac{1.44eV\bullet nm}{0.267nm}=-5.39eV$

(b) The net ionization energy is:

$E_{ion}=\left(\text{ionization energy of }Rb\right)-\left(\text{electron affinity of }Cl\right)$

$=4.18eV-3.62eV=0.56eV$

Neglecting the exclusion principle repulsion energy E_{ex},

dissociation energy $=-U(r_0)=5.39eV-0.56eV=4.83eV$

(c) Including exclusion principle repulsion,

dissociation energy $=4.37eV-U(r_0)=5.39eV-0.56eV-E_{ex}$

$E_{ex}=5.39eV-4.37eV-0.56eV=0.46eV$

9-9. For Kbr: $U_C=\dfrac{1.440eV\bullet nm}{0.282nm}+\left(4.34eV-3.36eV\right)=-4.13eV$

$E_d=3.94eV=\left|U_C+E_{ex}\right|=\left|-4.13eV+E_{ex}\right|$

$E_{ex}=0.19eV$

(Problem 9-9 continued)

$$\text{For } RbCl: U_C = \frac{1.440eV \bullet nm}{0.279nm} + (4.18eV - 3.62eV) = -4.60eV$$

$$E_d = 4.37eV = |U_C + E_{ex}| = |-4.60eV + E_{ex}|$$

$$E_{ex} = 0.23eV$$

9-13. $p_{ionic} = er_0 = (1.609 \times 10^{-19}C)(0.2345 \times 10^{-9}m)$ (Equation 9-3)

$$= 3.757 \times 10^{-29} C \bullet m, \text{ if purely ionic.}$$

The measured value should be:

$$p_{ionic}(measured) = 0.70 p_{ionic} = 0.70(3.757 \times 10^{-29} C \bullet m) = 2.630 \times 10^{-29} C \bullet m$$

9-17. $U = \alpha k^2 p_1^2 / r^2$ (Equation 9-10)

(a) Kinetic energy of $N_2 = 0.026eV$, so when $|U| = 0.026eV$ the bond will be broken.

$$0.026eV = \frac{(1.1 \times 10^{-37} m \bullet C^2 / N)(9 \times 10^9 N \bullet m^2 / C^2)^2 (6.46 \times 10^{-30} C \bullet m)^2}{r^6}$$

$$r^6 = \frac{(1.1 \times 10^{-37} m \bullet C^2 / N)(9 \times 10^9 N \bullet m^2 / C^2)^2 (6.46 \times 10^{-30} C \bullet m)^2}{0.026eV(1.60 \times 10^{-19} J / eV)} = 8.94 \times 10^{-56} m^6$$

$$r = 6.7 \times 10^{-10} m = 0.67nm$$

(b) $U \approx \dfrac{ke^2}{r} \rightarrow |U| = 0.026eV = \dfrac{1.440eV \bullet nm}{r} \rightarrow r \approx 55nm$

(c) H_2O-Ne bonds in the atmosphere would be very unlikely. The individual molecules will, on average, be about $4nm$ apart, but if a H_2O-Ne molecule should form, its $U \approx 0.003eV$ at $r = 0.95nm$, a typical (large) separation. Thus, a N_2 molecule with the average kinetic energy could easily dissociate the H_2O-Ne bond.

9-21. $E_{0r} = \dfrac{\hbar^2}{2I}$ (Equation 9-14) where $I = \dfrac{1}{2}mr_0^2$ for a symmetric molecule.

(Problem 9-21 continued)

$$E_{0r} = \frac{\hbar^2}{mr_0^2} = \frac{(\hbar c)^2}{mc^2 r_0^2} = \frac{(197.3\,eV\cdot nm)^2}{(16uc)^2\left(931.5\times10^6\,eV/uc^2\right)(0.121nm)^2} = 1.78\times10^{-4}\,eV$$

9-25. (a) $\mu = \dfrac{m_1 m_2}{m_1 + m_2} = \dfrac{(39.1u)(35.45u)}{39.1u + 35.45u} = 18.6u$

(b) $E_{0r} = \dfrac{\hbar^2}{2I}$ (Equation 9-14) $I = \mu r_0^2$

$$E_{0r} = \frac{\hbar^2}{2\mu r_0^2} = \frac{(\hbar c)^2}{2\mu c^2 r_0^2} \rightarrow r_0^2 = \frac{(\hbar c)^2}{2\mu c^2 E_V}$$

$$\therefore r_0 = \frac{\hbar c}{\left(2\mu c^2 E_V\right)^{1/2}} = \frac{197.3\,eV\cdot nm}{\left[2\left(10.6uc^2\right)\left(931.5\times10^6\,eV/uc^2\right)\left(1.43\times10^{-5}\,eV\right)\right]^{1/2}}$$

$r_0 = 0.280nm$

9-29. $E_{0r} = \dfrac{\hbar^2}{2I}$ where $I = \mu r_0^2$ (Equation 9-14)

For $K^{35}Cl$: $\mu = \dfrac{(39.102u)(34.969u)}{39.102u + 34.969u} = 18.46u$

For $K^{37}Cl$: $\mu = \dfrac{(39.102u)(34.966u)}{39.102u + 34.966u} = 19.00u$

$r_0 = 0.267nm$ for KCl.

$$E_{0r}\left(K^{35}Cl\right) = \frac{\left(1.055\times10^{-34}\,J\cdot s\right)^2}{2(18.46u)\left(1.66\times10^{-27}\,kg/u\right)\left(0.267\times10^{-9}\,m\right)^2}$$

$$= 2.55\times10^{-24}\,J = 1.59\times10^{-5}\,eV$$

$$E_{0r}\left(K^{37}Cl\right) = \frac{\left(1.055\times10^{-34}\,J\cdot s\right)^2}{2(19.00u)\left(1.66\times10^{-27}\,kg/u\right)\left(0.267\times10^{-9}\,m\right)^2}$$

$$= 2.48\times10^{-24}\,J = 1.55\times10^{-5}\,eV$$

$\Delta E_{0r} = 0.04\times10^{-5}\,eV$

9-33. $\dfrac{A_{21}}{B_{21}u(f)} = e^{hf/kT} - 1$ (Equation 9-42)

For the Hα line λ=656.1nm

At $T = 300K$, $\dfrac{hf}{kT} = \dfrac{hc}{\lambda kT} = \dfrac{1240eV \cdot nm}{(656.1nm)(8.62 \times 10^{-5} eV/K)(300K)} = 73.1$

$e^{hf/kT} - 1 = e^{73.1} - 1 \approx 5.5 \times 10^{31}$

Spontaneous emission is more probable by a very large factor!

9-37. $4mW = 4 \times 10^{-3} J/s$

$E = hc/\lambda = \dfrac{1240eV \cdot nm}{632.8nm} = 1.960eV \ per \ photon$

Number of photons $= \dfrac{4 \times 10^{-3} J/s}{(1.960eV)(1.60 \times 10^{-19} J/eV)} = 1.28 \times 10^{16} / s$

9-41.   (a) $\sin\theta = 1.22\lambda/D = 1.22(694.3 \times 10^{-9} m)/(0.01m) = 8.47 \times 10^{-5}$

$\therefore \ \theta = 8.47 \times 10^{-5} radians$

(b) $E_{photon} = hc/\lambda = 1240eV \cdot nm/694.3nm = 1.786eV / photon$

For $10^{18} photons/s$:

$E_{total} = (1.786eV / photon)(1.602 \times 10^{-19} J/eV)(10^{18} photons/s)$

$= 0.286J/s = 0.286W$

Area of spot A is: $A = \pi d^2/4 = \pi(8.47cm)^2/4$

and $E = E_{total}/A = 0.286W/\pi\left[(8.47cm)^2/4\right] = 5.08 \times 10^{-3} W/cm^2$

9-45.   (a) $E_3 = hc/\lambda = 1240eV \cdot nm/(0.86mm)(10^6 nm/mm) = 1.44 \times 10^{-3} eV$

$E_2 = 1240eV \cdot nm/(1.29mm)(10^6 nm/mm) = 9.61 \times 10^{-4} eV$

$E_1 = 1240eV \cdot nm/(2.59mm)(10^6 nm/mm) = 4.79 \times 10^{-4} eV$

(Problem 9-45 continued)

These are vibrational states, because they are equally spaced. Note the $v = 0$ state at the ½ level spacing.

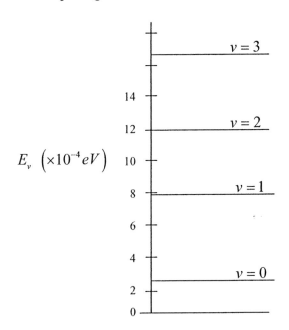

(b) Approximating the potential with a square well (at the bottom),

$$E_1 = 4.79 \times 10^{-4} eV = n^2 \frac{\pi^2}{2} \frac{\hbar^2}{mr_0^2}$$

$$r_0 = \left[\frac{(2^2 - 1^2)\pi^2 (1.055 \times 10^{-34} J \bullet s)^2}{2(28.01u)(1.66 \times 10^{-27} kg/u)(4.79 \times 10^{-4} eV)(1.60 \times 10^{-19} J/eV)} \right]^{1/2}$$

$$= 2.15 \times 10^{-10} m = 0.215 nm$$

9-49. $\mu(HCl) = 0.980u$ (See solution to Problem 9-47)

From Figure 9-29, the center of the gap is the characteristic oscillation frequency f:

$f = 8.65 \times 10^{13} Hz \rightarrow E = 0.36eV$ Thus, $f = \frac{1}{2\pi}\sqrt{\frac{K}{\mu}}$ or $K = (2\pi f)^2 \mu$

$$K = (2\pi)^2 (8.65 \times 10^{13} Hz)^2 (0.980u)(1.66 \times 10^{-27} kg/u) = 480 N/m$$

9-53. (a) $U(r) = -\dfrac{ke^2}{r} + E_{ex} + E_{ion}$ (Equation 9-1)

For $NaCl$, $E_d = 4.27eV$ and $r_0 = 0.236nm$ (Table 9-1).

$E_{ion} = E_{ion}(Na) + E_{aff}(Cl) = 5.14 = 3.62 = 1.52eV$ and $U(r_0) = -E_d = -4.27eV$

$E_{ex} = -4.27 + \dfrac{ke^2}{0.236} - 1.52 = 0.31eV$

(b) $E_{ex} = Ar^{-n} = 0.31eV$ (Equation 9-2)

Following Example 9-2, $\dfrac{ke^2}{r_0^2} = 25.85eV\,/\,nm = \dfrac{n}{r_0}\dfrac{A}{r_0^n} = \dfrac{n}{r_0}(0.31eV)$

Solving for n: $n = (25.85eV\,/\,nm)(0.236nm)/0.31eV = 19.7 \approx 20$

$A = (0.31eV)(0.236nm)^{20} = 8.9\times10^{-14}eV\cdot nm^{20}$

9-57.

#2 ; $n=1, v=1, \ell=1$; #3 $n=1, v=0, \ell=2$; #1 $n=1, v=0, \ell=0$

(a) $E(1) = \dfrac{1}{2}hf_{H_2} + \ell(\ell+1)E_{0r} = \dfrac{1}{2}hf_{H_2}$ since $\ell = 0$

$E(2) = \dfrac{3}{2}hf_{H_2} + 2E_{0r}$ since $\ell = 1$

$E(3) = \dfrac{1}{2}hf_{H_2} + 6E_{0r}$ since $\ell = 2$

[A] $E(2) - E(1) = \left(\dfrac{3}{2}hf_{H_2} + 2E_{0r}\right) - \dfrac{1}{2}hf_{H_2} = h(1.356\times10^{14}Hz)$

[B] $E(2) - E(3) = \left(\dfrac{3}{2}hf_{H_2} + 2E_{0r}\right) - \left(\dfrac{1}{2}hf_{H_2} + 6E_{0r}\right) = h(1.246\times10^{14}Hz)$

Re-writing [A] and [B] with $E_{0r} = \hbar^2/2I$:

[A1] $hf_{H_2} + \hbar^2/I = h(1.356\times10^{14}Hz)$

(Problem 9-57 continued)

$$[B1] \quad hf_{H_2} - 2\hbar^2/I = h\left(1.246 \times 10^{14}\,Hz\right)$$

Subtracting [B1] from [A1] and canceling an h from each term gives:

$$3h/4\pi^2 I = 0.110 \times 10^{14}\,Hz$$

$$I = \frac{3h}{4\pi^2 \left(0.110 \times 10^{14}\,Hz\right)} = 4.58 \times 10^{-48}\,kg \bullet m^2$$

(b) $I = \mu r_0^2$

For H_2 : $\mu = \dfrac{\left(1.007825\right)^2}{2\left(1.007825\right)} = 0.503912u$

$$r_0 = \left(I/\mu\right)^{1/2} = \left[\left(4.58 \times 10^{-48}\,kg \bullet m^2\right)/\left(0.5039u \times 1.66 \times 10^{-27}\,kg/u\right)\right]^{1/2}$$

$r_0 = 7.40 \times 10^{-11}\,m = 0.0740nm$ in agreement with Table 9-7.

Canceling an h from [B1] and substituting the value of I from (a) gives:

$$f_{H_2} = 2h/4\pi^2 I + 1.246 \times 10^{14}\,Hz$$

$f_{H_2} = 1.32 \times 10^{14}\,Hz$ also in agreement with Table 9-7.

Chapter 10 – Solid State Physics

10-1. $U(r_0) = -\alpha \dfrac{ke^2}{r_0}\left(1 - \dfrac{1}{n}\right)$ (Equation 10-6)

$$E = -U(r_0) = \alpha \dfrac{ke^2}{r_0}\left(1 - \dfrac{1}{n}\right)$$

$$1 - \dfrac{1}{n} = \dfrac{E_d r_0}{\alpha ke^2} = \dfrac{(741kJ/mol)(0.257nm)}{1.7476(1.44eV \cdot nm)} \times \dfrac{1eV/ion\ pair}{96.47kJ/mol} = 0.7844$$

$$n = \dfrac{1}{1 - 0.7844} = 4.64$$

10-5. Cohesive energy (LiBr)

$$= 788 \times 10^3 J/mol\left(\dfrac{1mol}{6.02 \times 10^{23} ion\ pairs}\right)\left(\dfrac{1eV}{1.60 \times 10^{-19}}\right) = 8.182ev/ion\ pair$$

$$= 4.09eV/atom$$

This is about 32% larger than the value in Table 10-1.

10-9. $U_{att} = -ke^2\left(\dfrac{2}{a} + \dfrac{2}{2a} - \dfrac{2}{3a} + \dfrac{2}{4a} + \dfrac{2}{5a} - \dfrac{2}{6a} + \cdots\right)$

$U_{att} = -ke^2\left(2 + 1 - \dfrac{2}{3} + \dfrac{1}{3} + \dfrac{2}{5} - \dfrac{1}{3} + \cdots\right)$

The quantity in parentheses is the Madelung constant α. The 35th term of the series (2/35) is approximately 1% of the total, where α = 4.18.

10-13. (a) $n = \rho N_A / M$ for 1 electron/atom

$$n = \dfrac{(10.5g/cm^3)(6.022 \times 10^{23}/mole)}{107.9g/mole} = 5.86 \times 10^{22}/cm^3$$

(b) $n = \dfrac{(19.3g/cm^3)(6.022 \times 10^{23}/mole)}{196.97g/mole} = 5.90 \times 10^{22}/cm^3$

Both agree with the values given in Table 10-3.

10-17. A long, thin wire can be considered one-dimensional.

$$E_F = \frac{h^2}{32m}\left(\frac{N}{L}\right)^2 = \frac{(hc)^2}{32mc^2}\left(\frac{N}{L}\right)^2 \quad \text{(Equation 10-15)}$$

For Mg: $N/L = \left(8.61\times10^{28}/m^2\right)^{1/3}$

$$E_F = \frac{\left(1240eV\bullet nm \times 10^{-9}\,m/nm\right)^2\left(8.61\times10^{28}/m^3\right)^{2/3}}{32\left(0.511\times10^6\,eV\right)} = 1.87eV$$

10-21. $\rho = \dfrac{m_e u_F}{ne^2\lambda}$ (Equation 10-25) $\qquad \lambda = \dfrac{m_e u_F}{ne^2\rho}$

(a) for Na: $\lambda = \dfrac{\left(9.11\times10^{-31}kg\right)\left(1.07\times10^6\,m/s\right)}{\left(2.65\times10^{28}\,m^{-3}\right)\left(1.609\times10^{-19}C\right)^2\left(4.2\times10^{-8}\Omega\bullet m\right)}$

$$= 3.42\times10^{-8}\,m = 34.2nm$$

(b) for Au: $\lambda = \dfrac{\left(9.11\times10^{-31}kg\right)\left(1.40\times10^6\,m/s\right)}{\left(5.90\times10^{28}\,m^{-3}\right)\left(1.609\times10^{-19}C\right)^2\left(2.04\times10^{-8}\Omega\bullet m\right)}$

$$= 4.14\times10^{-8}\,m = 41.4nm$$

(c) for Sn: $\lambda = \dfrac{\left(9.11\times10^{-31}kg\right)\left(1.90\times10^6\,m/s\right)}{\left(14.8\times10^{28}\,m^{-3}\right)\left(1.609\times10^{-19}C\right)^2\left(10.6\times10^{-8}\Omega\bullet m\right)}$

$$= 4.31\times10^{-8}\,m = 43.1nm$$

10-25. $P = \dfrac{\rho_+ - \rho_-}{\rho} = \dfrac{M}{\mu\rho} = \dfrac{\mu B}{kT}$ (Equation 10-35)

$$P = \frac{\left(9.285\times10^{-24}\,J/T\right)\left(2.0T\right)}{\left(1.38\times10^{-23}\,J/K\right)\left(200K\right)} = 6.7\times10^{-3}$$

10-29. (a) $E = hc/\lambda = 1240eV\bullet nm/\left(3.35\mu m \times 10^3\,nm/\mu m\right) = 0.37eV$

(b) $E = kT = 0.37eV \therefore T = 0.37eV/k = 0.37eV/8.617\times10^{-5}eV/K = 4300K$

10-33. Electron configuration of *Si*: $1s^2\, 2s^2\, 2p^6\, 3s^2\, 3p^2$

   (a) *Al* has a $3s^2\, 3p$ configuration outside the closed $n = 2$ shell (3 electrons), so a *p*-type semiconductor will result.

   (b) *P* has a $3s^2\, 3p^3$ configuration outside the closed $n = 2$ shell (5 electrons), so an *n*-type semiconductor results.

10-37. $I_{net} = I_0\left(e^{eV_b/kT} - 1\right)$ (Equation 10-49)

   (a) $e^{eV_b/kT} = 5$, so $eV_b/kT = \ln(5)$. Therefore,

$$V_b = \frac{kT\ln(5)}{e} = \frac{\left(1.38\times10^{-23}\,J/K\right)\left(200K\right)\ln(5)}{\left(1.60\times10^{-19}\,C\right)} = 0.0278V = 27.8mV$$

   (b) $e^{eV_b/kT} = 0.5$

$$eV_b/kT = \ln(0.5)$$

$$V_b = \frac{kT\ln(0.5)}{e} = \frac{\left(1.38\times10^{-23}\,J/K\right)\left(200K\right)\ln(0.5)}{\left(1.60\times10^{-19}\,C\right)} = -0.0120V = -12.0mV$$

10-41. $M^{\alpha}T_c = \text{constant}$ (Equation 10-55)

   First, we find the constant for *Pb* using the mass of natural *Pb* from Appendix A, T_c for *Pb* from Table 10-6, and α for *Pb* from Table 10-7.

$$\text{constant} = \left(207.19u\right)^{0.49}\left(7.196K\right) = 98.20$$

For $^{206}Pb : T_c = \text{constant}/M^{\alpha} = 98.20\big/\left(205.974u\right)^{0.49} = 7.217K$

For $^{207}Pb : T_c = \text{constant}/M^{\alpha} = 98.20\big/\left(206.976u\right)^{0.49} = 7.200K$

For $^{208}Pb : T_c = \text{constant}/M^{\alpha} = 98.20\big/\left(207.977u\right)^{0.49} = 7.183K$

10-45. $B_C(T)/B_C(0) = 1 - (T/T_C)^2$

    (a) $B_C(T)/B_C(0) = 0.1 = 1 - (T/T_C)^2$

$$(T/T_C)^2 = 1 - 0.1 = 0.9$$

$$T/T_C = 0.95$$

    (b) Similarly, for $B_C(T)/B_C(0) = 0.5$

$$T/T_c = 0.71$$

    (c) Similarly, for $B_C(T)/B_C(0) = 0.9$

$$T/T_c = 0.32$$

10-49. (a) $N = \int_0^{E_F} g(E)\,dE = \int_0^{E_F} AE^{1/2}\,dE = A(2/3)E^{3/2}\Big|_0^{E_F} = (2A/3)E_F^{3/2}$

    (b) $N' = \int_{E_F - kT}^{E_F} AE^{1/2}\,dE = (2A/3)\left[E_F^{3/2} - (E_F - kT)^{3/2}\right]$

$$= (2A/3)\left[E_F^{3/2} - E_F^{3/2}(1 - kT/E_F)^{3/2}\right]$$

Because $kT \ll E_F$ for most metals,

$$(1 - kT/E_F)^{3/2} \approx 1 - (3/2)kTE_F^{1/2}$$

$$N' = (2A/3)\left[E_F^{3/2} - E_F^{3/2} + (3/2)kTE_F^{1/2}\right] = AkTE_F^{1/2}$$

The fraction within kT of E_F is then $f = \dfrac{N'}{N} = \dfrac{AkTE_F^{1/2}}{(2A/3)E_F^{3/2}} = \dfrac{3kT}{2E_F}$

    (c) For Cu $E_F = 7.04eV$; at 300K, $f = \dfrac{3(0.02585eV)}{2(7.04eV)} = 0.0055$

10-53. $a_0 = \dfrac{\varepsilon_0 h^2}{\pi m_e e^2} = \dfrac{\kappa \varepsilon_0 h^2}{\pi (m_e)_{eff} e^2} = \dfrac{\kappa \hbar^2}{(m_e)_{eff} ke^2}$

silicon: $a_0 = \dfrac{12(1.055 \times 10^{-34}\, J{\cdot}s)^2}{0.2(9.11 \times 10^{-31}\, kg)(9 \times 10^9\, N{\cdot}m^2/C^2)(1.602 \times 10^{-19}\, C)^2}$

$= 3.17 \times 10^{-9}\, m = 3.17 nm$

This is about 14 times the lattice spacing in silicon (0.235nm)

germanium: : $a_0 = \dfrac{16(1.055 \times 10^{-34}\, J{\cdot}s)^2}{0.10(9.11 \times 10^{-31}\, kg)(9 \times 10^9\, N{\cdot}m^2/C^2)(1.602 \times 10^{-19}\, C)^2}$

$= 8.46 \times 10^{-9}\, m = 8.46 nm$

This is nearly 35 times the lattice spacing in germanium (0.243nm)

10-57. $U = -\alpha \dfrac{ke^2}{r_0}\left[\dfrac{r_0}{r} - \dfrac{1}{n}\left(\dfrac{r_0}{r} \right)^n \right]$ (Equation 10-5)

$F = -\dfrac{dU}{dr} = -Kr$ yields $K = \alpha \dfrac{(n-1)ke^2}{r_0^3}$

(a) For *NaCl*: $\alpha = 1.7476$, $n = 9.35$, and $r_0 = 0.282 nm$ and

$\mu = \dfrac{m(Na)\, m(Cl)}{m(Na) + m(Cl)} = \dfrac{(22.99u)(35.45u)}{(22.99u) + (35.45u)} = 13.95u$

$f = \dfrac{1}{2\pi}\sqrt{\dfrac{K}{M}} = \dfrac{1}{2\pi}\left[\dfrac{\alpha(n-1)ke^2}{(13.95u)\, r_0^3} \right]^{1/2}$

$= \dfrac{1}{2\pi}\left[\dfrac{(1.7476)(8.35)(9 \times 10^9\, N{\cdot}m^2/C^2)(1.60 \times 10^{-19}\, C)^2}{(13.95u \times 1.66 \times 10^{-27}\, kg/u)(0.282 \times 10^{-9}\, m)^3} \right]^{1/2} = 1.28 \times 10^{13}\, Hz$

(b) $\lambda = c/f = (.00 \times 10^8\, m/s)/1.28 \times 10^{13}\, Hz = 23.4 \mu m$

This is of the same order of magnitude as the wavelength of the infrared absorption bands in *NaCl*.

Chapter 11 – Nuclear Physics

11-1.

Isotope	Protons	Neutrons
^{18}F	9	9
^{25}Na	11	14
^{51}V	23	28
^{84}Kr	36	48
^{120}Te	52	68
^{148}Dy	66	82
^{175}W	74	101
^{222}Rn	86	136

11-5. The two proton spins would be antiparallel in the ground state with $S = 1/2 - 1/2 = 0$. So the deuteron spin would be due to the electron and equal to $1/2\hbar$. Similarly, the proton magnetic moments would add to zero and the deuteron's magnetic moment would be $1\mu_B$. From Table 11-1, the observed spin is $1\hbar$ (rather than $1/2\hbar$ found above) and the magnetic moment is $0.857\mu_N$, about 2000 times smaller than the value predicted by the proton-electron model.

11-9. $B = ZM_H c^2 + Nm_N c^2 - M_A c^2$ (Equation 11-11)

(a) $^9_4Be_5$ $B = 4(1.007825 uc^2) + 5(1.008665 uc^2) - 9.012182 uc^2$

   $= 0.062443 uc^2 = (0.062443 uc^2)(931.5 MeV/uc^2)$

   $= 58.2 MeV$

   $B/A = 58.2 MeV/9\ nucleons = 6.46 MeV/nucleon$

(b) $^{13}_6C_7$ $B = 6(1.007825 uc^2) + 7(1.008665 uc^2) - 13.003355 uc^2$

   $= 0.104250 uc^2 = (0.104250 uc^2)(931.5 MeV/uc^2)$

   $= 91.1 MeV$

   $B/A = 91.1 MeV/13\ nucleons = 7.47\ MeV/nucleon$

(Problem 11-9 continued)

(c) $^{57}_{26}Fe_{31}$ $\qquad B = 26\left(1.007825uc^2\right) + 31\left(1.008665uc^2\right) - 56.935396uc^2$

$\qquad\qquad\qquad\qquad = 0.536669uc^2 = \left(0.536669uc^2\right)\left(931.5MeV/uc^2\right)$

$\qquad\qquad\qquad\qquad = 499.9MeV$

$\qquad\qquad\qquad B/A = 499.9MeV/57\ nucleons = 8.77\ MeV/nucleon$

11-13. $R = \left(1.07 \pm 0.02\right)A^{1/3}\ fm$ (Equation 11-5) $\qquad R = 1.4A^{1/3}\ fm$ (Equation 11-7)

(a) ^{16}O: $\quad R = 1.07A^{1/3} = 2.70\ fm$ and $R = 1.4A^{1/3} = 3.53\ fm$

(b) ^{63}Cu: $\quad R = 1.07A^{1/3} = 4.26\ fm$ and $R = 1.4A^{1/3} = 5.57\ fm$

(c) ^{208}Pb: $\quad R = 1.07A^{1/3} = 6.34\ fm$ and $R = 1.4A^{1/3} = 8.30\ fm$

11-17. $R = R_0e^{-\lambda t} = R_0e^{-(\ln 2)t/t_{1/2}}$ (Equation 11-19)

(a) at $t = 0$: $\quad R = R_0 = 115.0\ decays/s$

$\qquad$ at $t = 2.25h$: $\quad R = 85.2\ decays/s$

$\qquad\qquad 85.2\ decays/s = \left(115.0\ decays/s\right)e^{-\lambda(2.25h)}$

$\qquad\qquad \left(85.2/115.0\right) = e^{-\lambda(2.25h)}$

$\qquad\qquad \ln\left(85.2/115.0\right) = -\lambda\left(2.25h\right)$

$\qquad\qquad\qquad \lambda = -\ln\left(85.2/115.0\right)/2.25h = 0.133h^{-1}$

$\qquad\qquad\qquad t_{1/2} = \ln 2/\lambda = \ln 2/0.133h^{-1} = 5.21h$

(b) $\left|\dfrac{dN}{dt}\right| = \lambda N \quad \rightarrow \quad \left|\dfrac{dN_0}{dt}\right| = R_0 = \lambda N_0$ $\quad$ (from Equation 11-17)

$\qquad N_0 = R_0/\lambda = \left(15.0\ atoms/s\right)/\left(0.133h^{-1}\right)\left(1h/3600s\right)$

$\qquad\qquad = 3.11 \times 10^6\ atoms$

11-21. ^{62}Cu is produced at a constant rate R_0, so the number of ^{62}Cu atoms present is:

$N = R_0 / \lambda \left(1 - e^{-\lambda t}\right)$ (from Equation 11-26). Assuming there were no ^{62}Cu atoms initially present. The maximum value N can have is $R_0 / \lambda = N_0$,

$$N = N_0 \left(1 - e^{-\lambda t}\right)$$
$$0.90 N_0 = N_0 \left(1 - e^{-t(\ln 2)/t_{1/2}}\right)$$
$$e^{-t(\ln 2)/t_{1/2}} = 1 - 0.90 = 0.10$$
$$-t \ln(2)/t_{1/2} = \ln(0.10)$$
$$t = -10 \ln(0.10)/\ln(2) = 33.2 \, min$$

11-25. $\log t_{1/2} = A E_\alpha^{-1/2} + B$ (Equation 11-30)

for $t_{1/2} = 10^{10} s$, $E_\alpha = 5.4 MeV$
for $t_{1/2} = 1s$, $E_\alpha = 7.0 MeV$ from Figure 11-16

$$\log(10^{10}) = A(5.4)^{-1/2} + B$$

(i) $10 = 0.4303 A + B$

$$\log(1) = A(7.0)^{-1/2} + B$$

(ii) $0 = 0.3780 A + B \rightarrow B = -0.3780 A$

Substituting (ii) into (i),

$$10 = 0.4303 A = 0.3780 A - 0.0523 A, \quad A = 191, \quad B = -0.3780 A = -72.2$$

11-29. $^{67}Ga \xrightarrow{E.C.} {}^{67}Zn + v_e$

$$Q = M\left({}^{67}Ga\right)c^2 - M\left({}^{67}Zn\right)c^2$$
$$= 66.9282uc^2 - 66.972129uc^2$$
$$= 0.001075uc^2 \left(931.50 MeV/uc^2\right) = 1.00 MeV$$

11-33. $^8Be \rightarrow 2\alpha$

$$Q = M\left(^8Be\right)c^2 - M\left(^4He\right)c^2$$

$$= 8.005304uc^2 - 2(4.002602)uc^2$$

$$= 0.000100uc^2\left(931.50MeV/uc^2\right) = 0.093MeV = 93keV$$

11-37. The range R of a force mediated by an exchange particle of mass m is:

$R = \hbar/mc$ (Equation 11-50)

$mc^2 = \hbar c/R = 197.3MeV \cdot fm/0.25fm = 789MeV$

$m = 789MeV/c^2$

11-41. ^{36}S, ^{53}Mn, ^{82}Ge, ^{88}Sr, ^{94}Ru, ^{131}In, ^{145}Eu

11-45. $^{30}_{14}Si$ $j = 0$

$^{37}_{17}Cl$ $j = 3/2$

$^{55}_{27}Co$ $j = 7/2$

$^{90}_{40}Zr$ $j = 0$

$^{107}_{49}In$ $j = 9/2$

11-49. (a) $^{12}C\left(\alpha, p\right)^{15}N$

$$Q = M\left(^{12}C\right)c^2 + M\left(^4He\right)c^2 - M\left(^{15}N\right)c^2 - m_p c^2$$

$$= 12.000000uc^2 + 4.002602uc^2 - 15.000108uc^2 - 1.007825uc^2$$

$$= -0.005331uc^2\left(931.5MeV/uc^2\right) = -4.97MeV$$

(b) $^{16}O\left(d, p\right)^{17}O$

$$Q = M\left(^{16}O\right)c^2 + M\left(^2H\right)c^2 - M\left(^{17}O\right)c^2 - m_p c^2$$

$$= 15.994915uc^2 + 2.014102uc^2 - 16.999132uc^2 - 1.007825uc^2$$

$$= 0.002060uc^2\left(931.5MeV/uc^2\right) = 1.92MeV$$

11-53. $\dfrac{Q}{c^2} = m_p + m_n - m_d$

$\qquad = 1.007276u + 1.008665u - 2.013553u$

$\qquad = 0.002388u \quad$ (See Table 11-1.)

$Q = \left(0.002388u\right)\left(931.5 MeV / uc^2\right)c^2 = 224 MeV$

11-57. $500 MW = \left(500\dfrac{J}{s}\right)\left(\dfrac{1 MeV}{1.60\times10^{-13}J}\right)\left(\dfrac{1\,fusion}{17.6 MeV}\right) = 1.78\times10^{14}\,fusions\,/\,s$

Each fusion requires one ^2H atom (and one ^3H atom; see Equation 11-67) so ^2H must be provided at the rate of $1.78\times10^{14}\,atoms\,/\,s$.

11-61. $\rho(H_2O) = 1000 kg / m^3$, so

(a) $1000 kg$: $\quad \dfrac{10^6 g \left(6.02\times10^{23}\,molecules\,H_2O/mol\right)\left(2H/molecule\right)\left(0.00015\,^2H\right)}{18.02 g/mol}$

$\qquad = 1.00\times10^{25}\ ^2H\ atoms$

Each fusion releases $5.49 MeV$.

$\qquad$ Energy release $= \left(1.00\times10^{25}\right)\left(5.49 MeV\right) = 5.49\times10^{25}\,MeV$

$\qquad\qquad = \left(5.49\times10^{25}\,MeV\right)\left(1.60\times10^{-13}\,J/MeV\right) = 8.78\times10^{12}\,J$

(b) Energy used/person (in 1999) $= 3.58\times10^{20}\,J\big/5.9\times10^{9}\,people$

$\qquad\qquad = 6.07\times10^{10}\,J\,/\,person{\bullet}y$

Energy used per person per hour $= 6.07\times10^{10}\,J\,/\,person{\bullet}y\times\dfrac{1y}{8760h}$

$\qquad\qquad = 6.93\times10^{6}\,J\,/\,person{\bullet}h$

At that rate the deuterium fusion in $1m^3$ of water would last the "typical" person

$\qquad \dfrac{8.78\times10^{12}\,J}{6.93\times10^{6}\,J\,/\,person{\bullet}h} = 1.27\times10^{6}\,h \approx 145 y$

11-65.
$$t = \frac{t_{1/2}}{\ln(2)} \ln(1 + N_D / N_P) \qquad \text{(Equation 11-92)}$$

$$t_{1/2}\left(^{87}Rb\right) = 4.88 \times 10^{10} \, y \quad \text{and} \quad N_P / N_D = 36.5$$

$$t = \frac{4.88 \times 10^{10} \, y}{\ln(2)} \ln\left[1 + (1/36.5)\right] = 1.90 \times 10^{9} \, y$$

11-69. (a) $N\left(^{12}C^{+3}\right) = \dfrac{\left(12 \times 10^{-6} \, C/s\right)\left(10\,min\right)\left(60s / min\right)}{3\left(1.60 \times 10^{-19} \, C\right)} = 1.50 \times 10^{16}$

$^{14}C/^{12}C$ ratio $= 1500/1.50 \times 10^{16} = 10^{-13}$

(b) mass $^{12}C = \dfrac{\left(1.50 \times 10^{15} \, atoms / min\right)\left(75\,min\right)\left(12\right)\left(1.66 \times 10^{-27} \, kg\right)}{0.015}$

$$= 1.49 \times 10^{-7} \, kg = 1.49 \times 10^{-4} \, g = 0.149 mg$$

(c) The $^{14}C/^{12}C$ ratio in living C is 1.35×10^{-12}.

$$\frac{\text{sample } ^{14}C/^{12}C}{\text{living } ^{14}C/^{12}C} = \frac{10^{-13}}{1.35 \times 10^{-12}} = \frac{0.10}{1.35} = \left(\frac{1}{2}\right)^{n}$$

where $n = $ # of half-lives elapsed. Rewriting as (see Example 11-28)

$$2^{n} = \frac{1.35}{0.10} = 13.5$$

$$n\ln(2) = \ln(13.5) \qquad \therefore n = \ln(13.5)/\ln(2) = 3.75$$

age of sample $= 3.75 t_{1/2} = 3.75(5730y) = 2.15 \times 10^{4} \, y$

11-73. For one proton, consider the nucleus as a sphere of charge e and charge density $\rho_c = 3e/4\pi R^3$. The work done in assembling the sphere, i.e., bringing charged shell dq up to r, is: $dU_c = k\rho_c \dfrac{4\pi r^3}{3}\left(\rho_c 4\pi r^2 dr\right)\dfrac{1}{r}$ and integrating from 0 to R yields:

$$U_c = \frac{k\rho_c^2 16\pi^2 R^5}{15} = \frac{3}{5}\frac{ke^2}{R}$$

For two protons, the coulomb repulsive energy is twice U_c, or $6ke^2/5R$.

11-77. (a) $\Gamma = \hbar/\tau = 6.582 \times 10^{-16} eV \cdot s/0.13 \times 10^{-9} s = 5.06 \times 10^{-6} eV$

(b) $E_r = \dfrac{(hf)^2}{2Mc^2} = \dfrac{(0.12939 MeV)^2}{2M(^{191}I)c^2}$ (Equation 11-47)

$$= 4.71 \times 10^{-8} MeV = 4.71 \times 10^{-2} eV$$

(c) (See Section 1-5)

The relativistic Doppler shift Δf for either receding or approaching is:

$$\frac{\Delta f}{f_0} \approx \beta = \frac{v}{c} \qquad h\Delta f = \Gamma \qquad hf_0 = E$$

$$\frac{\Gamma}{E} = \frac{v}{c} \rightarrow v = \frac{\Gamma c}{E} = (5.06 \times 10^{-6} eV)(3.00 \times 10^8 m/s) = 0.0117 m/s = 1.17 cm/s$$

11-81. (a) $R = R_0 A^{1/3}$ where $R_0 = 1.2 f$ (Equation 11-3)

$$R(^{141}Ba) = (1.2 fm)(10^{-15} m/fm)(141)^{1/3} = 6.24 \times 10^{-15} m$$

$$R(^{92}Kr) = (1.2 fm)(10^{-15} m/fm)(92)^{1/3} = 5.42 \times 10^{-15} m$$

(b) $V = kq_1 q_2 / r = \dfrac{(8.998 \times 10^9 N \cdot m^2/C^2)(56)(1.60 \times 10^{-19} C)(36)(1.60 \times 10^{-19} C)}{(6.24 \times 10^{-15} m + 5.42 \times 10^{-15} m)(1.60 \times 10^{-19} J/eV)}$

$$= 2.49 \times 10^8 eV = 249 MeV$$

This value is about 40% larger than the measured value.

11-85. The number of ^{87}Sr atoms present at any time is equal to the number of ^{87}Rb nuclei that have decayed, because ^{87}Sr is stable.

$$N(Sr) = N_0(Rb) - N(Rb) \rightarrow N(Sr)/N(Rb) = N_0(Rb)/N(Rb) - 1 N$$

$$N(Sr)/N(Rb) = 0.010$$

$$N_0(Rb)/N(Rb) = N(Sr)/N(Rb) + 1 = 1.010$$

and also

$$N(Rb)/N_0(Rb) = e^{-(\ln 2)t/t_{1/2}} = 1/1.010$$

(Problem 11-85 continued)

$$\frac{-(\ln 2)t}{t_{1/2}} = \ln(1/1.010)$$

$$t = -t_{1/2}\ln(1/1.010)/\ln(2) = -\left(4.9\times 10^{10}\,y\right)\ln(1/1.010)/\ln(2) = 7.03\times 10^8\,y$$

11-89. (a) $\Delta\lambda \leq 2hc/Mc^2$

$$\Delta E \approx \frac{hc\Delta\lambda}{\lambda^2} = \frac{(hc)^2}{\lambda^2}\frac{\Delta\lambda}{hc} = \frac{E^2\Delta\lambda}{hc}$$

$$E_p = \Delta E \leq \frac{E^2}{hc}\frac{2hc}{Mc^2} = \frac{2E^2}{Mc^2}$$

$$E^2 \geq Mc^2 E_p/2 \quad \rightarrow \quad E \geq \left(Mc^2 E_p/2\right)^{1/2}$$

$$\Delta E = E_f - E_i = E_i\left(1-\frac{4mM}{(M+m)^2}\right) - E_i = -E_i\left(\frac{4mM}{(M+m)^2}\right)$$

$$\frac{-\Delta E}{E_i} = \frac{4mM}{(M+m)^2} = \frac{4m/M}{(1+m/M)^2} \quad \text{which is Equation 11-82 in More section.}$$

(b) $E = \left[(5.7MeV)(938.28MeV)/2\right]^{1/2} = 51.7MeV$

(c)

The neutron moves at v_L in the lab, so the *CM* moves at $v = v_L m_N/(m_N + M)$ toward the right and the ^{14}N velocity in the *CM* system is v to the left before collision and v to the right after collision for an elastic collision. Thus, the energy of the nitrogen nucleus in the lab after the collision is:

$$E\left(^{14}N\right) = \frac{1}{2}M\left(2v\right)^2 = 2Mv^2 = 2M\left(\frac{mv_L}{m+M}\right)^2$$

$$= \frac{2Mm\left(mv_L^2\right)}{(m+M)^2} = \frac{4Mm}{(m+M)^2}\left(\frac{1}{2}mv_L^2\right)$$

(Problem 11-89 continued)

$$= \frac{4(14.003074u)(1.008665u)}{(1.008665u + 14.003074u)^2}(5.7MeV)$$

$$= 1.43 \ MeV$$

(d) $E \geq \left[(14.003074uc^2)(931.5MeV \, / \, uc^2)(1.43MeV) \, / \, 2 \right]^{1/2} = 96.5MeV$

11-93. (a)

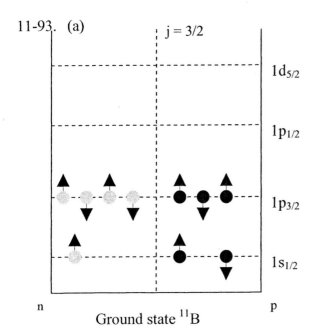

Ground state ^{11}B

(b)

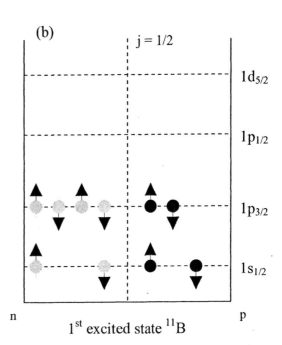

1st excited state ^{11}B

(c)

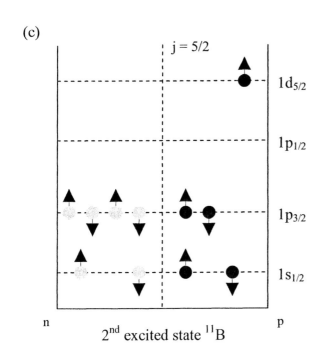

2nd excited state ^{11}B

(Problem 11-93 continued)

(d)

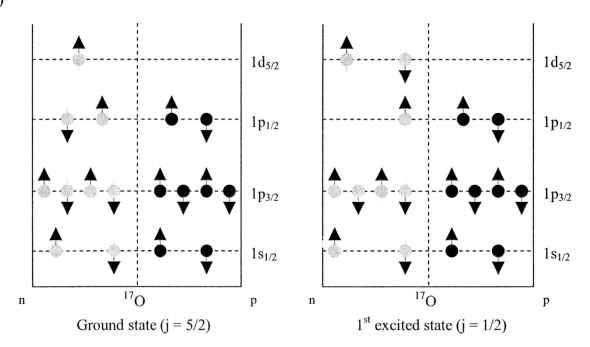

Ground state (j = 5/2)   1ˢᵗ excited state (j = 1/2)

(e)

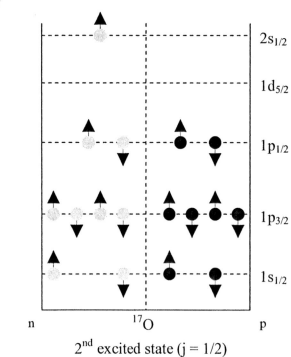

2ⁿᵈ excited state (j = 1/2)

11-97. $\dfrac{dN}{dt} = R_p = -\lambda N$ (Equation 11-17)

(a) $\lambda N = R_p - \lambda N_0 e^{-\lambda t} = R_p - R_p e^{-\lambda t} = R_p \left(1 - e^{-\lambda t}\right)$

$N = \left(R_p / \lambda\right)\left(1 - e^{-\lambda t}\right)$ (from Equation 11-17)

At $t = 0$, $N(0) = 0$. For large t, $N(t) \to R_p / \lambda$, its maximum value

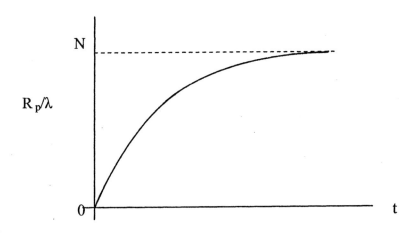

(b) For $dN / dt \approx 0$

$R_p = \lambda N \to N = R_p / \lambda = R_p / \left(\ln 2 / t_{1/2}\right)$

$N = 100 s^{-1} / \left(\ln 2 / 10\, min\right) = \left(100 s^{-1}\right)\left(60 s / min\right) / \left(\ln 2 / 10\, min\right)$

$= 8.66 \times 10^4 \quad {}^{62}Cu$ nuclei

11-101. (a) The number N of generations is: $N = \dfrac{5s}{0.08 s / gen} = 62.5$ generations

Percentage increase in energy production $= \dfrac{R(N) - R(0)}{R(0)} \times 100$

$= \left[\dfrac{R(N)}{R(0)} - 1\right] \times 100$ where $R(N) / R(0) = k^N$ (from Example 11-22 in More section)

$= \left(k^N - 1\right) \times 100 = \left(1.005^{62.5} - 1\right) \times 100 = 137\%$

(Problem 11-101 continued)

(b) Because $k \propto neutron\ flux$, the fractional change in flux necessary is equal to the fractional change in k:

$$\frac{k-1}{k} = \frac{1.005-1}{1.005} = 0.00498$$

Chapter 12 – Particle Physics

12-1. (a) Because the two pions are initially at rest, the net momentum of the system is zero, both before and after annihilation. For the momentum of the system to be zero after the interaction, the momentum of the two photons must be equal in magnitude and opposite in direction, i.e., their momentum vectors must add to zero. Because the photon energy is $E = pc$, their energies are also equal.

(b) The energy of each photon equals the rest energy of a π^+ or a π^-.

$E = m_\pi c^2 = 139.6 MeV$ (from Table 12-3)

(c) $E = hf = hc / \lambda$ Thus, $\lambda = \dfrac{hc}{E} = \dfrac{1240 MeV \bullet fm}{139.6 MeV} = 8.88\, fm$

12-5. (a) $^{32}P \rightarrow {}^{32}S + e^-$ assuming no neutrino

$Q = M\left({}^{32}P\right)c^2 - M\left({}^{32}S\right)c^2$ (electron's mass is included in that of ^{32}S)

$= 31.973908 uc^2 - 31.972071 uc^2$

$= \left(0.001837 uc^2\right)\left(931.5 MeV / uc^2\right) = 1.711 MeV$

To a good approximation, the electron has all of the kinetic energy

$E_k \approx Q = 1.711 MeV$

(b) In the absence of a neutrino, the ^{32}S and the electron have equal and opposite momenta. The momentum of the electron is given by:

$\left(pc\right)^2 = E^2 - \left(m_e c^2\right)^2$ (Equation 2-32)

$= \left(E_k + m_e c^2\right)^2 - \left(m_e c^2\right)^2$

$= \left(Q + m_e c^2\right)^2 - \left(m_e c^2\right)^2 = Q^2 + 2Q m_e c^2$

The kinetic energy of the ^{32}S is then:

$E_k = \dfrac{p^2}{2M} = \dfrac{\left(pc\right)^2}{2Mc^2} = \dfrac{Q^2 + 2Q m_e c^2}{2M\left({}^{32}S\right)c^2}$

(Problem 12-5 continued)

$$= \frac{(1.711MeV)^2 + 2(1.711MeV)(0.511MeV)}{2(31.972071uc^2)(931.5MeV/uc^2)}$$

$$= 7.85 \times 10^{-5} MeV = 78.5eV$$

(c) As noted above, the momenta of the electron and ^{32}S are equal in magnitude and opposite direction.

$$(pc)^2 = Q^2 + 2Qm_ec^2 = (1.711MeV)^2 + 2(1.711MeV)(0.511MeV)$$

$$p = \left[(1.711MeV)^2 + 2(1.711MeV)(0.511MeV)\right]^{1/2} \Big/ c$$

$$= 2.16MeV/c$$

12-9.   (a) Weak interaction

(b) Electromagnetic interaction

(c) Strong interaction

(d) Weak interaction

12-13. For neutrino mass $m = 0$, travel time to Earth is $t = d/c$, where $d = 170,000c \cdot y$.   For neutrinos with mass $m \neq 0$, $t' = d/v = d/\beta c$, where $\beta = v/c$.

$$\Delta t = t' - t = \frac{d}{c}\left(\frac{1}{\beta} - 1\right) = \frac{d}{c}\left(\frac{1-\beta}{\beta}\right)$$

$$\gamma = \frac{1}{\sqrt{1-\beta^2}} \quad \text{(Equation 1-19)}$$

$$\gamma^2 = \frac{1}{1-\beta^2} = \frac{1}{(1-\beta)(1+\beta)}$$

$$1-\beta = \frac{1}{\gamma^2(1+\beta)} \approx \frac{1}{2\gamma^2} \quad \text{since } \beta \approx 1$$

Substituting into Δt,

(Problem 12-13 continued)

$$\Delta t \approx \frac{d}{c}\left(\frac{1}{2\gamma^2}\right) \qquad E = \gamma mc^2 \rightarrow \gamma^2 = \left(E/mc^2\right)^2 \quad \text{(Equation 2-10)}$$

$$\Delta t \approx \frac{d}{2c}\left(\frac{mc^2}{E}\right)^2$$

$$mc^2 = \left(\frac{(\Delta t)\,2cE^2}{d}\right)^{1/2} = \left(\frac{2\Delta t E^2}{d/c}\right)^{1/2}$$

$$= \left[\frac{2(12.5s)\left(10\times10^6\,eV\right)^2}{(170{,}000\,c\cdot y/c)\left(3.16\times10^7\,s/y\right)}\right]^{1/2} = 21.6eV$$

$$m \approx 22eV/c^2$$

12-17. (a) $m_p c^2 < \left(m_n + m_e\right)c^2$ Conservation of energy and lepton number are violated.

(b) $m_n c^2 < \left(m_p + m_\pi\right)c^2$ Conservation of energy is violated.

(c) Total momentum in the center of mass system is zero, so two photons (minimum) must be emitted. Conservation of linear momentum is violated.

(d) No conservation laws are violated. This reaction, $p\bar{p}$ annihilation, occurs.

(e) Lepton number before interaction is +1; that after interaction is −1. Conservation of lepton number is violated.

(f) Baryon number is +1 before the decay; after the decay the baryon number is zero. Conservation of baryon number is violated.

12-21. (a) $\pi^- \rightarrow e^- + \gamma$ Electron lepton number changes from 0 to 1; violates conservation of electron lepton number.

(b) $\pi^0 \rightarrow e^- + e^+ + v_e + \bar{v}_e$ Allowed by conservation laws, but decay into two photons via electromagnetic interaction is more likely.

(c) $\pi^+ \rightarrow e^- + e^+ + \mu^+ + v_\mu$ Allowed by conservation laws but decay without the electrons is more likely.

(Problem 12-21 continued)

(d) $\Lambda^0 \to \pi^+ + \pi^-$ Baryon number changes from 1 to 0; violates conservation of baryon number. Also violates conservation of angular momentum, which changes from 1/2 to 0.

(e) $n \to p + e^- + \bar{\nu}_e$ Allowed by conservation laws. This is the way the neutron decays.

12-25. Listed below are the baryon number, electric charge, strangeness, and hadron identity of the various quark combinations from Table 12-8 and Figure 12-21.

	Quark Structure	Baryon Number	Electric Charge (e)	Strangeness	Hadron
(a)	uud	+1	+1	0	p
(b)	udd	+1	0	0	n
(c)	uuu	+1	+2	0	Δ^{++}
(d)	uss	+1	0	−2	Ξ^0
(e)	dss	+1	−1	−2	Ξ^-
(f)	suu	+1	+1	−1	Σ^+
(g)	sdd	+1	−1	−1	Σ^-

Note that 3-quark combinations are baryons.

12-29. The +2 charge can result from either a *uuu*, *ccc*, or *ttt* quark configuration. Of these, only the *uuu* structure also has zero strangeness, charm, topness, and bottomness. (From Table 12-5.)

12-33. $n \to p + \pi^-$

$$Q = m_n c^2 - m_p c^2 - m_\pi c^2$$

$$= \left(939.6 - 938.3 - 139.6\right) MeV$$

$$= -138.3 MeV$$

This decay does not conserve energy.

12-37. (a) Being a meson, the D^+ is constructed of a quark-antiquark pair. The only combination with *charge* = +1, *charm* = +1 and *strangeness* = 0 is the $c\bar{d}$. (See Table 12-5.)

(b) The D^-, antiparticle of the D^+, has the quark structure $\bar{c}d$.

12-41. (a) $p \rightarrow e^+ + \Lambda^0 + v_e$

$$Q = \left(m_p c^2 - M\left(\Lambda^0\right)c^2 - m_e c^2 \right) MeV$$

$$= \left(938.3 - 1116 - 0.511 \right) MeV = -178 MeV$$

Energy is not conserved.

(b) $p \rightarrow \pi^+ + \gamma$

Spin (angular momentum) $\dfrac{1}{2} \rightarrow 0 + 1 = 1$. Angular momentum is not conserved.

(c) $p \rightarrow \pi^+ + K^0$

Spin (angular momentum) $\dfrac{1}{2} \rightarrow 0 + 0 = 0$. Angular momentum is not conserved.

12-45. (a) The final products (p, γ, e^-, neutrinos) are all stable.

(b) $\Xi^0 \rightarrow p + e^- + \overline{v}_e + \overline{v}_\mu + v_\mu$

(c) Conservation of charge: $0 \rightarrow +1 - 1 + 0 + 0 + 0 = 0$

Conservation of baryon number: $1 \rightarrow 1 + 0 + 0 + 0 + 0 = 1$

Conservation of lepton number:

(i) for electrons: $0 \rightarrow 0 + 1 - 1 + 0 + 0 = 0$

(ii) for muons: $0 \rightarrow 0 + 0 + 0 - 1 + 1 = 0$

Conservation of strangeness: $-2 \rightarrow 0 + 0 + 0 + 0 + 0 = 0$

Even though the chain has $\Delta S = +2$, no individual reaction in the chain exceeds

$\Delta S = +1$, so they can proceed via the weak interaction.

(d) No, because energy is not conserved.

12-49. (a) $\Lambda^0 \rightarrow p + \pi^-$

Energy: $1116 MeV - \left(938 + 140 \right) MeV = 38 MeV$ conserved.

Electric charge: $0 \rightarrow +1 - 1 = 0$ conserved.

Baryon number: $1 \rightarrow 1 + 0 = 1$ conserved.

Lepton number: $0 \rightarrow 0 + 0 = 0$ conserved.

(b) $\Sigma^- \rightarrow n + p^-$

Energy: $1197 MeV - \left(940 + 938 \right) MeV = -681 MeV$ not conserved.

(Problem 12-49 continued)

Electric charge: $-1 \rightarrow 0-1 = -1$ conserved.

Baryon number: $1 \rightarrow 1-1 = 0$ not conserved.

Lepton number: $0 \rightarrow 0+0 = 0$ conserved.

This reaction is not allowed (energy and baryon conservation violated).

(c) $\mu^- \rightarrow e^- + \bar{\nu}_e + \nu_\mu$

Energy: $105.6 MeV - 0.511 MeV = 105.1 MeV$ conserved.

Electric charge: $-1 \rightarrow -1+0+0 = -1$ conserved.

Baryon number: $0 \rightarrow 0+0+0 = 0$ conserved.

Lepton number:

 (i) electrons: $0 \rightarrow 1-1+0 = 0$ conserved.

 (ii) muons: $1 \rightarrow 0+0+1 = 1$ conserved.

12-53. $\Sigma^0 \rightarrow \Lambda^0 + \gamma$

(a) E_T for decay products is the rest energy of the Σ^0, $1193 MeV$.

(b) The rest energy of $\Lambda^0 = 116 MeV$, so $E_\gamma = 1193 MeV - 1116 MeV = 77 MeV$

and $p_\gamma = E_\gamma / c = 77 MeV / c$.

(c) The Σ^0 decays at rest, so the momentum of the Λ^0 equals in magnitude that of the photon.

$$E_{kin}\left(\Lambda^0\right) = p_\Lambda^2 / 2M\left(\Lambda\right) = \left(77 MeV/c\right)^2 / \left[2\left(1116 MeV/c^2\right)\right]$$

$$= 2.66 MeV \quad \text{small compared to } E_\gamma$$

(d) A better estimate of E_γ and p_γ are then $E_\gamma = 77 MeV - 2.66 MeV = 74.3 MeV$ and

$p_\gamma = 74.3 MeV / c.$

Chapter 13 – Astrophysics and Cosmology

13-1.

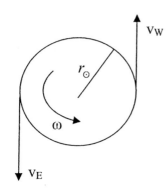

$|v_W - v_E| = 4 km/s$. Assuming Sun's rotation to be uniform, so that $v_W = -v_E$, then $|v_W| = |v_E| = 2 km/s$. Because $v = 2\pi/T$, $v_E = 2\pi r_\odot / T$ or

$$T = \frac{2\pi r_\odot}{v_E} = \frac{2\pi(6.96 \times 10^5 km)}{2 km/s} = 2.19 \times 10^6 s = 25.3\, days$$

13-5. Observed mass (average) $\approx 1\, H\, atom/m^3 = 1.67 \times 10^{-27} kg/m^3 = 4\%$ of total mass

$\therefore$ missing mass dark matter $= (22/4) \times 1.67 \times 10^{-27} kg/m^3 = 9.19 \times 10^{-27} kg/m^3$

$500 v/cm^3 = 500 \times 10^6 v/m^3$, so the mass of each v would be

$$= \frac{9.19 \times 10^{-27} kg/m^3}{500 \times 10^6 v/m^3} = 1.84 \times 10^{-35} kg \quad \text{or} \quad m_v = \frac{1.84 \times 10^{-35} kg}{1.60 \times 10^{-19} J/eV} \times \frac{c^2}{c^2}\left(\frac{m^2}{s^2}\right) = 10.4 eV/c^2$$

13-9. $L = 4\pi r^2 f$ $\qquad m_1 - m_2 = 2.5\log(f_1/f_2)$

Thus, $L_p = 4\pi r_p^2 f_p$ and $L_B = 4\pi r_B^2 f_B$ and $L_p = L_B$

$\therefore r_p^2 f_p = r_B^2 f_B \rightarrow r_B^2 = r_p^2 (f_p/f_B)$

$\log(f_p/f_B) = \dfrac{1.16 - 0.41}{2.5} = 0.30 \rightarrow f_p/f_B = 2.00$

Because $r_p = 12 pc$, $r_B = r_p (f_p/f_B)^{1/2} = 12\sqrt{2} = 17.0 pc$

13-13.

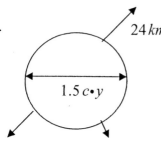

24 km/s

(a) $r = \dfrac{1.5 c \cdot y}{2}$; assuming constant expansion rate,

$$\text{Age of Shell} = \frac{1.5 c \cdot y/2}{2.4 \times 10^4 m/s} = 2.95 \times 10^{11} s = 9400 y$$

(b)

(Problem 13-13 continued)

$$R \propto M \to R = \alpha M \qquad T_e \propto M^{1/2} \to T_e = \beta M^{1/2} \qquad L \propto M^4 \to L = \gamma M^4$$

$$\therefore \alpha = R_\odot / M_\odot, \qquad \beta = T_{e\odot} / M_\odot^{1/2}, \qquad \gamma = L_\odot / M_\odot^4$$

$$R_{star} = \frac{R_\odot}{M_\odot} m_{star}, \qquad T_{e\,star} = \frac{T_{e\odot}}{M_\odot^{1/2}} M_{star}^{1/2}, \qquad L_{star} = \frac{L_\odot}{M_\odot^4} M_{star}^4$$

Using either the T_e or L relations, $R_{star} = \dfrac{M_{star}}{M_\odot} R_\odot = \left(\dfrac{T_{e\,star}}{T_{e\odot}}\right)^2 = R_\odot = (1.4)^2 R_\odot = 1.96 R_\odot$

or $R_{star} = \left(\dfrac{L_{star}}{L_\odot}\right)^{1/2} = 1.86 R_\odot$

13-17. $v = 72,000 \, km/s$.

(a) $v = Hr \to r = \dfrac{v}{H} = \dfrac{72,000 km/s}{22 km/s/10^6 c\cdot y} = 3.27 \times 10^9 c\cdot y$

(b) From Equation 13-29 the maximum age of the galaxy is:

$$1/H = 4.30 \times 10^{17} s = 1.36 \times 10^{10} y$$

$$1/H = r/v \to \Delta(1/H) = \Delta r/v \quad \therefore \quad \frac{\Delta(1/H)}{(1/H)} = \frac{\Delta r}{r} = 10\%$$

so the maximum age will also be in error by 10%.

13-21. $\rho(\text{Planck time}) = \dfrac{m_{pl}}{\ell_{pl}^3} = \dfrac{5.5 \times 10^{-8} kg}{(10^{-35})^3 m^3} = 5.5 \times 10^{97} kg/m^3$

$$\rho(\text{proton}) = \frac{1.67 \times 10^{-27} kg}{(10^{-15})^3 m^3} = 1.67 \times 10^{18} kg/m^3$$

$$\rho(\text{osmium}) = 2.45 \times 10^4 kg/m^3$$

13-25.

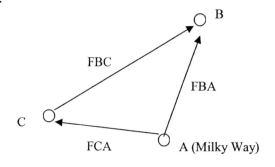

If Hubble's law applies in A, then

$$v_{BA} = Hr_{BA}, \quad v_{CA} = Hr_{CA}.$$

From mechanics,

$$v_{BC} = v_{BA} - v_{CA} = H(r_{BA} - r_{CA}) = Hr_{BC}$$

and Hubble's lab applies in C, as well, and by extension in all other galaxies.

13-29. $M_\odot = 1.99 \times 10^{30} kg.$

(a) When first formed, mass of $H = 0.7M_\odot$, $m(^1H) = 1.007825u \times 1.66 \times 10^{-27} kg/u$, thus

$$\text{number of H atoms} = \frac{0.7 \times M_\odot}{1.007825u \times 1.66 \times 10^{-27} kg/u} = 8.33 \times 10^{56}$$

(b) If all $H \to He$; $4\,^1H \to\,^4He + 26.72eV$. The number of He atoms

$$\text{produced} = \frac{8.33 \times 10^6}{4}.$$

$$\text{Total energy produced} = \frac{8.33 \times 10^6}{4} \times 26.72 MeV = 5.56 \times 10^{57} MeV = 8.89 \times 10^{44} J$$

(c) 23% of max possible $= 0.23 \times 8.89 \times 10^{44} J$

$$t_L = \frac{0.23 \times 8.89 \times 10^{44}}{L_\odot} = 5.53 \times 10^{17} s = 1.7 \times 10^{10} y \qquad (L_\odot = 3.85 \times 10^{26} W)$$

13-33.

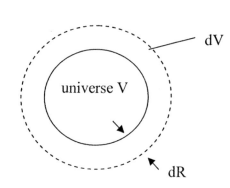

$$H = \frac{22km/s}{10^6 c\bullet y}$$

Current average density $= 1H\ atom/m^3$

$$V = \frac{4}{3}\pi R^3 \to dV = 4\pi R^2 dR$$

(Problem 13-33 continued)

The current expansion rate at R is:

$$v = HR = \frac{22 km/s}{10^6 \, c \bullet y} \times 10^{10} \, c \bullet y = 22 \times 10^4 \, km/s = 22 \times 10^7 \, m/s$$

$$\therefore \quad dR = 22 \times 10^7 \, m/s \times 3.16 \times 10^7 \, s/y \times \frac{10^6 \, y}{10^6 \, y}$$

$$dV = 4\pi R^2 \, dR = 4\pi \times \left(10^{10}\right)^2 \left(9.45 \times 10^{15} \, m/c \bullet y\right)^2 \times 22 \times 10^7 \, m/s \times 3.16 \times 10^7 \, s/y \times \frac{10^6 \, y}{10^6 \, y}$$

$$= \frac{7.78 \times 10^{74} \, m^3}{10^6 \, c \bullet y} = \frac{\text{\# of H atoms}}{10^6 \, c \bullet y} \quad \text{to be added}$$

Current volume $V = \frac{4}{3}\pi \left(10^{10}\right)^3 = 8.4 \times 10^{77} \, m^3$

$$\therefore \quad \text{"new" H atoms} = \frac{7.78 \times 10^{74} \, atoms/10^6 \, c \bullet y}{8.4 \times 10^{77} \, m^3} \approx 0.001 \text{ "new" H } atoms/m^3 \bullet 10^6 \, c \bullet y \, ; \quad \text{no}$$

13-37. (a)

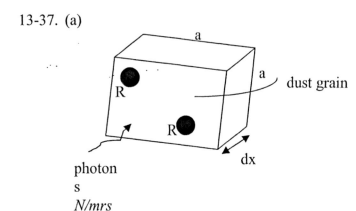

photon
s
N/mrs

$n = grains/cm^3$

total scattering area $= \pi R^2 n a^2 \, dx$

which is $\dfrac{\pi R^2 a^2 n \, dx}{a^2} = \pi R^2 n \, dx$ of the

total area = fraction scattered = dN/N

$$\int_{N_0}^{N} \frac{dN}{N} = -n\pi R^2 \int_0^d dx \quad \text{or} \quad N = N_0 e^{-n\pi R^2 d}$$

From those photons that scatter at $x = 0$ (N_0), those that have not scattered again after traveling some distance $x = L$ is $N_L = N_0 e^{-n\pi R^2 L}$. The average value of L ($= d_0$) is given by:

(Problem 13-37 continued)

$$d_0 = \frac{\int_0^\infty L \frac{dN_L}{dL} dL}{\int_0^\infty \frac{dN_L}{dL} dL} = \frac{1}{n\pi R^2} \qquad \left(\text{Note: } \frac{dN_L}{dL} = -n\pi R^2 N_0 e^{-n\pi R^2 L} \right)$$

(b) $I = I_0 e^{-d/d_0}$ near the Sun $d_0 \approx 3000 \, c \bullet y$ $R = 10^{-5} \, cm$

$$\therefore \quad 3000 c \bullet y \times 9.45 \times 10^{17} \, cm/c \bullet y = \frac{1}{n\pi \left(10^{-5}\right)^2} \qquad \therefore \quad n = 1.1 \times 10^{-12} \, / cm^3$$

(c) $\rho_{grains} = 2 \, gm/cm^3$

$$\therefore \quad \frac{m_{grains}}{cm^3 \text{ of space}} = 2 \times \frac{4}{3} \pi \left(10^{-5}\right)^3 \times 1.1 \times 10^{-12} \, / cm^3 = 9.41 \times 10^{-27} \, gm/cm^3$$

$$\therefore \quad \text{mass in } 300 c \bullet y = \frac{9.41 \times 10^{-27} \, gm/cm^3}{M_\odot} \times \left(9.45 \times 10^{17} \, cm/c \bullet y\right)^3 \times 300$$

$$= 0.0012 \quad \left(\approx 0.1\% M_\odot\right)$$